D1228578

LION OF WHITE HALL

THE LIFE OF CASSIUS M. CLAY

The Aged Lion: Cassius M. Clay at eighty-four. Taken November 13, 1894, by Lexington photographer Isaac C. Jenks. From an original negative in the collection of J. Winston Coleman, Jr.

LION OF WHITE HALL

THE LIFE OF CASSIUS M. CLAY

David L. Smiley

GLOUCESTER, MASS.
PETER SMITH
1969

Published by the University of Wisconsin Press
430 Sterling Court, Madison 6, Wisconsin

Reprinted, 1969 by Permission of
David L. Smiley

TO HELEN AND KAY

PREFACE

FOR forty years, when the American people were engaged in heated disputes over slavery, the Civil War, and the pangs of Reconstruction, Cassius M. Clay of Kentucky was a dramatic and controversial participant in the conflict. Even in his youth the tales of his colorful deeds had become legendary, and in his eccentric old age the myths grew until they overshadowed the bizarre personality of the man himself. There was sufficient factual basis for the fanciful tales. Clay was widely feared as a fighter who wielded a mighty knife with which he disembowelled his political opponents and carved off ears and excavated eyes in gory combat.

The notorious fighting, however, was but a part of the strange story of Cassius M. Clay. A cousin of the more restrained Henry Clay, Cassius was a man of paradoxes. No simple description could convey the complexity of his personality. He was an abolitionist who, almost alone, remained in the heart of slave territory. He fought under the banner of humanitarian and liberal reform and expected to win personal rewards for it. He was an industrial promoter in the land of the plantation, and a pioneer Republican in a Democratic stronghold. He was a rough-and-tumble fighter who lived past his fortieth birthday solely because of his brawny physique and his ready violence; yet he was noted for his polite conversation, his literary polish, and his admiration for the arts. He was a diplomat with a bowie knife in his hand, and a scion of landed aristocracy who championed the cause of the common man.

Complex and paradoxical though he was, his career revealed a singleness of purpose. His objective was to attain political power and public office, and his every act was calculated to accomplish it. In the end he was unsuccessful: his career is a study in political failure. It is also illustrative of the extremes to which an ambitious man would go in his efforts to gratify his ambitions. And because of the nature of the fight Clay waged, the story is a chapter in the pervasive influence of the race problem in the South.

I am under deep obligation to many persons and institutions for their help in my study of Cassius M. Clay: to the officials and librarians of the Public Library of Lexington, Kentucky, the University of Kentucky, Transylvania University, Berea College, Eastern Kentucky State College at Richmond, Western Kentucky State College at Bowling Green, the Kentucky State Historical Society, Lincoln Memorial University in Harrogate, Tennessee, Wake Forest College, the University of North Carolina, Duke University, the Henry E. Huntington Library, the Massachusetts Historical Society, the Filson Club, the Philosophical and Historical Society of Ohio in Cincinnati, the University of Rochester, the Historical Society of Pennsylvania, the Wisconsin State Historical Society, and the Library of Congress, all of whom made available their manuscripts and printed materials.

Furthermore, I appreciate the assistance given me by Colonel Eldon W. Downs of the United States Air Force, Dr. J. T. Dorris of Eastern Kentucky State College, Dr. Larry Gara of Grove City College, Dr. Sam Ross of Sacramento State College, and by Mr. Charles R. Staples and Mr. J. Winston

Coleman, Jr., both of Lexington, all of whom provided helpful advice and notes from their files.

My thanks also go to Mr. Warfield Bennett and to Miss Helen Bennett, both of Richmond, Kentucky, grandchildren of Cassius M. Clay, who graciously encouraged me in my study and made available family traditions and portraits; to Mr. Charles T. Dudley and Mrs. Jane Clay, both of Richmond, and to Mrs. Leonora R. Bergman, of Lexington, for personal reminiscences; and to Mr. Cassius M. Clay, of Paris, Kentucky, for permitting me access to the valuable Brutus J. Clay Papers.

In addition, I am deeply indebted to Dr. William B. Hesseltine, of the University of Wisconsin, who suggested the subject and whose sympathetic guidance was of inestimable value. My wife has rendered unusual services as critic and proofreader. I also wish to thank the Trustees and Administration of Wake Forest College for granting me a leave of absence.

Portions of this work have appeared in the *Journal of Mississippi History*, *Register of the Kentucky State Historical Society*, *Lincoln Herald*, *Filson Club History Quarterly*, *Bulletin of the Historical and Philosophical Society of Ohio*, and the *Journal of Negro History*, and I wish to thank the editors for permission to use it here.

D. L. S.

Winston-Salem, North Carolina
March, 1961

CONTENTS

ILLUSTRATIONS

LION
OF
WHITE HALL

THE LIFE OF CASSIUS M. CLAY

CHAPTER I

THE TRIALS OF A YOUNGEST SON

YOUNG Cassius Clay had told a lie, and his mother threatened him with a whipping. Having previously experienced his mother's punishing hand, the boy chose flight rather than submission. The house servants, fully sympathetic with Mrs. Clay's zeal for truth, gladly joined the chase. Cassius soon saw that they would overtake him, and spying a pile of stones, he determined to fight his pursuers. There, he said later, "I took my stand and made things lively." [1]

In that action, Cassius Marcellus Clay provided a preview of his career. Many times in the course of his life, his enemies —some moved by a hatred for untruths, some by distaste at the course he took, and others by the sheer zest of the hunt —closed in upon him. In each case he chose to fight, and each time he made things lively. Reviewing a life punctuated by violence, excitement, and romance, Clay recalled that fighting characterized all of its nine decades. Possessed of a stubborn spirit and an indomitable will, he would surrender no issue, however insignificant.

At important crises of his life he resorted to violence: he fought a mock duel with a hickory cane over his first marriage, and prepared private artillery to defend his second; he fought a colleague in the Kentucky legislature with his fists; with pistols and bowie knives he disputed political issues with inimical Kentucky neighbors. Later, as United States Minister to the Court of the Czars during Lincoln's administration, he won the admiration of Russian noblemen with his warlike propensities. Nor did he limit his militant spirit to weighty matters: he issued defiant challenges to personal combat over such quarrels as the genealogy of a Shorthorn bull or the origins of Kentucky bluegrass. With pungent pen, homely weapons, and unassisted brute strength, he defended his contentions.

Such active belligerence won widespread attention. Clay's temper, and his ability to win fights, became legendary. "Naturally pugnacious," a contemporary said of him, "he would fight the wind did it blow from the South side when he wanted it to blow from the North." Clay's fights added to Kentucky's history a colorful record of physical combat; and the names of his opponents constituted a roll call of the most prominent citizens of the Bluegrass State.[2]

But Cassius Clay was not content merely to fight individuals, for he was more than a hot-tempered duellist. The most important conflict of Clay's career, the one that lay at the root of his other clashes and brought him to the attention of the American people, was his attack upon slavery. At a time when the southern creed demanded unswerving devotion to the gospel of slavery, Clay was a heretic. When men of the Old South were opposing the industrialists of a new age, he criticized their established ways. Basing his attack upon economic and political arguments rather than humanitarian considerations, Clay denounced the labor system that characterized the plantation and its agrarian economy.

Though slave-owners denounced him and fought him, they could not drive him out of the state. "Having the full courage of his convictions," an admirer eulogized, "he took his life in his hands and bearded the pro-slavery lion in his den." In taking such a stand, Clay renounced his own birthright. He had inherited a large Bluegrass estate and a proud name, and he had married into one of Kentucky's first families. He was therefore closely allied with the state's social and political leaders, and in the beginning he received their blessing. As soon as he was old enough to qualify, he served in the state legislature, and he was on the threshold of a promising career. Fame, wealth, and a comfortable life—all were his if he would but become a satisfied member of his class and cater to its prejudices. But he would not. Doggedly, uncompromisingly, he held to a course that meant political ruin in the immediate future, but promised—he hoped—eventual reward. He refused to surrender merely because his words were unpopular; instead, he repudiated the attitudes his contemporaries cherished, and they cut him off from their favors. Clay, however, had larger aspirations. Pursuing a minority course and flying in the face of his fellows made him a nationally known figure. Carefully nurturing his reputation as a man of courage, a Republican in a Democratic stronghold, a martyr to the cause of human liberty and free speech, Clay played his cards for high stakes. His stubborn political revolt caused him to lose fortune, family, and success. His devotion to high and noble principles, he complained on lecture trips to the North, had consigned him to a dreary desert of political failure. And in 1860, when his political allies finally attained control of the Presidency, Clay offered his twenty-year exile from public office as a reason for being awarded a responsible post in the new administration.

Cassius M. Clay was not merely an erratic nonconformist, seeking publicity and reveling in notoriety. Behind his apparently mad rebellion against his people, his region, and his class lay reasons carefully calculated and schemes deliberately concocted. Basically, Cassius Clay's career was a search for political office and power. His calculations and his schemes eventually failed, and he did not live in the White House. But in the effort to achieve his ambitions he became one of the most important minor figures of the nineteenth century, and he left a distinctive mark upon the pages of American history.

Perhaps, in his own fashion, Cassius Clay was carrying out the ambitious drives that had brought his father to Kentucky and made him seek wealth and prestige on a frontier which was developing into a settled community. Cassius owed much to Green Clay, though the directions in which ambition pushed the father and the son were different indeed. Green Clay's successful career was possible because he hewed to the line of his community's conventions; the son's controversial search for political power defied the mores of his society.

Green Clay was but one of the hundreds who emigrated to Kentucky in the eighteenth century, but unlike many another pioneer, he was able to wring success from the Dark and Bloody Ground. Leaving Virginia in 1780 as a poor boy, Green Clay learned, as many of his fellows did not, that the secret of success in the wilderness of "Kentake" was land, well chosen and properly exploited. And the easiest way for a newcomer to obtain land, he discovered, was to clear it out for others.

Eager and intelligent, young Green Clay soon mastered the technique of the compass and chain and became an enterprising surveyor. With his knowledge of the country and his untiring zeal, he amassed large blocks of land from his commissions, which according to the custom of the early days often amounted to one-half of the property cleared. In the brief span of fifteen years the boy who had arrived penniless became one of Kentucky's leading citizens.[3] In 1788, when Kentucky was still a county of Virginia, he served in the Richmond convention that considered ratifying the federal constitution, and he voted with the majority of the Kentucky delegation against ratification. The next year he represented Madison County (a new county in the Kentucky territory) in the Virginia legislature. It is unlikely that he would have attained such eminence had he remained in the older community.[4]

But he demonstrated his abilities in interests other than political. Years later, Cassius said of his father that "his life was one rather of business than anything else; and here he passed all his contemporaries in the West." The son's appraisal was not far wrong. In 1792, Green Clay became a commissioner in a scheme to operate a toll road from the falls of the Great Kanawha River, in western Virginia, to Lexington, Kentucky; the route lay through his lands, and he controlled the Kentucky River ferry which would serve travelers on the road.

Thus, though Green Clay concentrated upon the accumulation of landed estates, he did not lose sight of the advantages of a diversified economy. He operated a distillery in Madison County and another across the river in Fayette County. At Estill Springs he built a resort which became popular in early Kentucky, and he built and rented taverns as outlets for the product of his distilleries. By 1795, as a commander in the local militia, justice of the peace, enterpriser in commercial and industrial ventures, and owner of choice Kentucky lands, Green Clay had become an outstanding citizen of his state. He set an enviable record for those who would later bear his name.[5]

During his years of active fortune-hunting, however, he gave no evidence that there would be any heirs to either his economic interests or his respected position. He devoted his entire attention to business, and not until he had assured his success did he marry. His wife, Sally Lewis, daughter of Kentucky pioneers Thomas Lewis and Elizabeth Payne, was nearly twenty years younger, and much more polished, than the rough, practical Clay.

After their wedding, on March 14, 1795, Green took his nineteen-year-old bride to their new home—a rough-hewn log cabin with a dirt floor on the uplands of Tate's Creek, not far from the Kentucky River. It was a good location, carefully chosen from the thousands of acres Green Clay commanded, but it was far from the settlements at Boonesborough, and even farther from Harrodstown. There, in the wilderness, the Clays made their home. The family was soon increased by children—at first all daughters. Finally, after three girls had been born, Sally gave birth to a son, who was named Sidney Payne. A few years later, in 1808, another son was born and was named Brutus Junius.[6]

The family had outgrown the log cabin, and Sally had long been agitating for a more pretentious dwelling. Green Clay built a handsome brick building to satisfy her and to crown his successes as an opulent pioneer. He named it White Hall. There, in the master bedroom, on October 19, 1810, Sally Clay presented her husband a third son. Continuing his fondness for classical names, the father named the baby Cassius Marcellus.[7]

From the beginning of his life the youngest son demonstrated a precocious belligerence. Typically, one of his earliest memories was of a fight won by a stratagem. One day, playing with George, a young slave boy on the White Hall plantation, he injured his playmate. "Mars' Cash," complained the Negro boy, "you would not treat me so if you had not marster and mistress to back you."

Eagerly, Cassius arose to the implied challenge. "Well, George," he answered, "I can whip you myself."

"If you won't tell, we'll see," George responded, and the two boys sought a secluded spot for the encounter. Young Cassius knew that in size and in strength he was no match for the gangling Negro, so he quickly worked out a plan: he would select a favorable field for the battle. He went to the side of the White Hall lawn where there was a steep descent. Stone for the house had been removed from the hillside in horizontal layers, leaving a level bench now covered with leaves. Cassius took his stand facing downhill, and George unthinkingly opposed him. Striking George unexpectedly in the face, Cassius sent him staggering downhill. Then, advancing to the edge of the shelf, he found himself taller than his opponent, and he had the advantage of fighting downward. He showered blows upon the Negro's unprotected head and soon had the victory. Through years of conflict, Cassius never forgot the thrill of conquering an adversary physically stronger than himself. His career as a fighter had begun, and self-confidence and the heady thrill of victory drove him to innumerable combats.[8]

There were other fights in Cassius Clay's early years which gave evidence of his virile belligerence, and there were also incidents which indicated an undiplomatic streak of stubbornness within his character. One such event concerned his father's wine chest. Green Clay gave strict orders that none of the children was to touch it, and every morning, as an object lesson in temperance, he would take a drink of bourbon and then make a face at the boys as though it tasted horrible. But hardheaded Cassius would not allow any teaching to take the place of experience; he sampled the contents of the bottle for himself. Though he afterwards found little fascination in bourbon, he had learned that his father's act was a sham.

Another memory from his early days suggested even more clearly his stubborn nature. His father had imported a merino buck sheep and had tied him to a tree in the yard. Cassius teased the animal until it lowered its head and prepared for battle. The youngster would not let even an aroused buck sheep intimidate him. He lowered his head, in imitation of the buck, and invited a trial of hardness of heads. His father, coming on the scene at that dramatic moment, knocked Cassius out of the sheep's path just in time. Later, jokingly, Cassius remarked that friends said his father had taken needless precautions, for his head would have proved too much for the sheep. But, figuratively at least, it was true: the incident reflected both his "try-anything-once" trait of foolhardiness and his stubborn courage. In years to come, opposition as hardy and as dangerous as a buck sheep faced him, and without a qualm he lowered his head and charged away.

Life for young Cassius was more than fighting and butting at sheep. There were embryonic love affairs with barefoot neighborhood girls. There were hunting and fishing expeditions, activities which held a lifelong attraction for Cassius. There were many hours at lessons, over which Mrs. Clay presided. Green Clay, whose formal education was limited but whose worldly wisdom was extensive, taught his sons valuable lessons in self-reliance. When Cassius was twelve years old, his father sent him to Cincinnati to pay taxes on some Ohio lands he owned. In 1823, with few roads and rough travel, the journey of over a hundred miles was a dangerous undertaking for one so young. Cassius realized afterwards that Green Clay intended the trip as part of his practical education.[9]

Cassius' father could also finance the best formal schooling available. After finishing four neighborhood common schools, Cassius and his brother Brutus enrolled under a private tutor, Joshua Fry, the most popular teacher in central Kentucky. The school was held on Fry's farm, on the banks of the Dix River, where Cassius happily spent leisure hours fishing and enjoying the out-of-doors. While he was not engaged in such pleasant activities he studied Latin, rhetoric, and philosophy—the education of the polished young aristocrat of the day.[10]

After several years of Fry's instruction, the Clay brothers had received as much formal education as was generally acquired. Brutus was ready to begin his long and successful career as a planter, but Cassius wished to continue his studies. He faced life from the viewpoint of a youngest son who had to compete with successful elder brothers, both of whom had political and social aspirations. His numerous fights had set him apart from the amiable Brutus, but now he proposed to differentiate himself still further. He went on to seek more education.

Green Clay gave his encouragement and support. In 1827 Cassius enrolled in the Jesuit College of St. Joseph at Bardstown, Kentucky. There his only study was French, and he never learned it well. That was unfortunate, for French was the diplomatic language of the western world, and when he later became a member of the foreign service he would regret that he had not applied himself more energetically to French. But his stay among the adherents of Roman Catholicism may have contributed to his tolerance in religious matters. He was, indeed, tolerant to the point of indifference. Henceforth his religious philosophy tended to humanistic deism, rather than orthodox Christianity.

With the distasteful lessons in language and his own rebellion against orthodoxy, Cassius did not enjoy his stay with the fathers, who tried in vain to bridle his belligerent temper. The high-spirited youth resisted the close supervision and wrote his brother Brutus that he craved the "pleasure of being unrestrained." That pleasure he would seek throughout his life, even while serving as a foreign minister. But at St. Joseph's his rebellious nature manifested itself, as it would later, in physical combat with his fellows. One of the students, with a reputation as a bully, was torturing one of the younger boys, when Cassius (as he recounted many years later) "sprang upon a bench, and hit him a stinging blow upon the nose, which caused the blood to fly in all directions." His attack, he reported, cured the fellow of his "evil ways, and made me quite a hero." [11]

While he was struggling with the French language and with his schoolmates, Cassius was, in addition, worried about his father's health. For some time Green Clay had been suffering with a disease diagnosed as cancer of the face. The son wanted to leave school to be with his parents. "Is not mother most worn out with fatigue," he wondered, "waiting on father so long?" On April 15, 1828, concerned about affairs at home, he left St. Joseph's and returned to White Hall to help his overburdened mother in the sickroom.[12]

He found Green Clay, now seventy-one years old, on his deathbed, growing weaker every day. Throughout his father's months of ebbing strength, Cassius was the old man's nurse and constant companion. He helped Green Clay arrange the family business, and watched while he wrote and revised his will. An unusual document, the last testament of Green Clay bequeathed to the youngest son extensive properties, which determined the immediate course of his career and influenced all of his life. His inheritance gave Cassius assistance in his efforts to catch up with his older brothers.

To Cassius, Green Clay left in trust "the tract of land on which I live containing about 2000 acres." Should Cassius die without issue, however, the White Hall estate would revert to Brutus. The father also entrusted specified slaves to members of the family; Cassius received a total of seventeen of his father's Negroes. In addition to the home tract and the slaves, the youngest son shared in the division of his father's other holdings. "My lands and land claims below the Tennessee River, . . . about 40 or 50,000 acres," according to the old man's directions, were left in fee to each of the six children, but for all except the two oldest sons the property was entrusted. Green Clay also created a contingent fund to provide for the needs of his children, and out of that fund provided for "the schooling and support of my son Cassius until he arrive at full age; then he shall take possession of his estate." It was the father's intention that none of his children should ever suffer want.

Green Clay's will assured Cassius of as much schooling as he desired; moreover, it provided him with a handsome estate. The settlement also made him a slave-owner, but not for years did the fact embarrass him. Perhaps that part of the will most disconcerting to him in future years was the restriction that his inheritance of the White Hall estate was not in fee simple but was entailed. He could not profitably dispose of it if he should desire to emigrate from Kentucky. It may have been an insignificant clause in his father's will which kept Cassius Clay in Kentucky and made him a leader in the *southern* antislavery movement.

Furthermore, by establishing him as a shareholder in vast land claims and in commercial and industrial ventures, Green Clay determined the direction in which Cassius' early career would lead. Having a responsible position as one of his father's legal executors, and being himself one of the property trustees, Cassius was thrown before he was twenty into the world of business affairs. The experience soon acquainted him with the delicate alliance between economic and political forces.

His father's will also affected his future life. As a property-owner, he would never advocate the destruction of property rights (in slaves) by any other than legal means. Moreover, his position as a member of the landed aristocracy of Kentucky provided him firm support for political discussion. No one would ever have grounds to say of him, as men would say of Hinton R. Helper of North Carolina, that he was a "poor-white" who was jealous of the more fortunate, or that he was a "have-not" who desired to have. And forty years later, when he moved among Russian noblemen as American Minister, he felt at ease because his life had been spent among the polite and the well-to-do at home. To this extent Green Clay's fortune influenced the career of his youngest son.

Having provided for his family, the father calmly awaited the end. Stoically he had endured the pains of his disease and watched as his health slowly faded. At last, on the night of October 31, 1828, the old man called Cassius to his bedside.

"I have just seen death come in that door," he muttered faintly, gesturing in the direction of the family graveyard. These were his last words.[13]

For nearly eighteen years Cassius had been under the influence and tutelage of his father. Green Clay, wealthy, respected, and justly proud of his self-made success, left with the son practical lessons in the code of the frontier gentleman and businessman. He had attempted to pass along his experience-bought knowledge by means of terse, homely epigrams, and more than fifty years later Cassius could still remember his father's advice.

Green Clay wanted his sons to be self-sufficient and cautious in their dealings with other men—lessons particularly applicable to a diplomatic career. "Never tell any one your business," the father warned; and "Never set your name on the right-hand side of the writing" (meaning that they should never sign as security for another's note—a piece of advice Cassius would ruefully remember in years to come). "Gambling and toping I warn you against; . . . keep out of the hands of the doctor and the sheriff." He advised his sons not to sell on credit: "My property is worth more on the farm, or in the storeroom," he said, "than in the pockets of spendthrifts." He cautioned Cassius to be suspicious of strangers and never to trust anyone but himself. "Although you think you can speak in confidence to a friend," Green counselled, "that friend will betray you in all probability at some future day when he can wound you deepest." To these apothegms Cassius listened and learned. A simple philosophy emphasizing independence, caution, and self-reliance, based upon a pessimistic view of human nature, was the creed that the father bequeathed to his son.[14]

Cassius' mother also exerted an influence upon the personality of the growing boy, and he was conscious of her part in his character development. "At all times," he said, summing up the sources of his nature, "the mother, being both parent and teacher, mostly forms the character." Sally Clay was a deeply religious woman, a Calvinistic Baptist who felt the hand of her God in every incident of her life. She believed it to be her duty, along with teaching the primer, to convert her sons to her own deterministic faith. Many years after Cassius had left her fireside, Sally's letters to him contained imploring religious appeals. "I have been trying to serve the Lord upward of forty years, I then believed I could be saved by the imputed righteousness of Christ, and I rejoiced with joy unspeakable and full of glory," she wrote in 1849. "I know you can't understand me except you experience the same, but the Lord is able to show you what a helpless sinner you are and the vanity of all earthly things." When Cassius joined a church, it was a Baptist congregation. But he resisted his mother's efforts to make him a deeply religious person like herself; he preferred the rationalistic deism of his father. Though Sally Clay's primary concern was for the souls of her sons, she did not neglect more worldly matters. She taught the boys to observe the rules of polite etiquette as she knew them. Under her training Cassius learned the generous chivalry of the southern gentleman.[15]

But that was not all Mrs. Clay transmitted to her youngest son. She endowed him with a share of her rebellious nature. Much of Cassius' belligerence might be traced to his mother. Like many another Bible-quoting Puritan, Sally Clay was dangerous when aroused, and she encouraged her sons to stand up for their rights. It was she who urged Cassius to fight when a mob threatened him, and other relatives were advising surrender. Despite her age, she came loyally to nurse him when he was wounded in fights with political enemies. Throughout his career she stood by Cassius, and she would sanction no dishonorable retreat. Cassius owed much in his personality to his strong-willed parents. But the most important trait of his character—his driving ambition—Cassius gained through observation. Like his father before him, he wanted to become influential, to be respected as a leader among men. For Green Clay's youngest son, the road to respect and influence led through the exciting arena of politics. Cassius Clay's ambition was to hold high office.

To prepare himself for a career in public life, Cassius continued his formal education. A few months after his father died, he entered Transylvania University in Lexington, a flourishing community across the Kentucky River from his home. From its imposing courthouse to its gracious homes, including that of the statesman Henry Clay, Lexington was a pleasant city and a center of Kentucky business, particularly Bluegrass agriculture. It was the home of a great commerce, though it was fifteen miles from navigable water, and it provided congenial surroundings for manufacturers and artisans. Lexington was also fast becoming the most important center of the Kentucky slave trade, and on Cheapside, near the courthouse, slave auctions were frequently held.

The cultural center of the city was its college. Transylvania University, founded in 1780, was the oldest college west of the mountains and contributed to Lexington's reputation as the "Athens of the West." [16] The university and the growing city stimulated the imagination and the intellect of Cassius Clay.

When he entered the Junior Class in January, 1829, the college boasted three buildings set upon a flat, grassy campus on the north edge of the town. Two of the structures were low and rambling, but the main building was the pride of the school and of the town. Constructed of red brick, it housed the library, classrooms, the "philosophical laboratory," and administrative offices. A third floor contained living quarters for students, and there Cassius roomed.

Already accustomed to student life, he fitted easily into the college routine. Of the sports which were an important part of the young men's activities, he was particularly fond.

"I had ever been devoted to athletic sports—riding on horseback, boxing, hunting, fishing, gunning, jumping, scuffling, wrestling, playing 'base-ball,' bandy, 'football,' and all that," he recalled later. Many years after his student days at Transylvania, he remembered the excitement of a close game of "bandy" on the campus west of the main building.[17]

But there was more to his college memories than sports. Looking back on his college years, Clay felt that he had been a success as a student. How assiduous a scholar he was may be open to question: the librarian's journal shows that, in his two years at Transylvania, Cassius, though he was entitled to take out of the university library at least one book a week, checked out only two volumes, both in his first month at the school, January, 1829. Of the studies offered at Transylvania, ranging from surveying to rhetoric, from evidences of Christianity to chemistry "with experiments," the most important to the ambitious youth was oratory, and he spent one afternoon a week in directed speaking. He also joined a "philosophical society" and took an active part in its program.[18]

While at Transylvania, too, Cassius heard some of Kentucky's noted orators of the day: Henry Clay, the "Gallant Harry of the West"; Robert J. Breckinridge, a Lexington Presbyterian minister who advocated gradual emancipation of slaves; Robert Wickliffe, the "Old Duke," who later became the most important of Cassius' political opponents; and William T. Barry, Postmaster General in Andrew Jackson's cabinet. Clay's interest in oratory and public debate indicated his growing interest in public affairs.[19]

At the university he met fellow students who afterwards became leaders in their communities or who played a part in his own life. One of his classmates was N. L. Rice, who became famous as debater against the Virginian Alexander Campbell over a theological issue. Others were Montgomery Blair, who later played a significant role in Missouri's Civil

War drama and served in Abraham Lincoln's cabinet; Robert Wickliffe, Jr., the "Young Duke," against whom Cassius was to run for office and fight a duel in years to come; and James B. Clay, a son of Henry Clay.

Among the experiences at Transylvania that influenced Clay's developing personality, two others stand out. One, with which he later claimed a close connection, was a fire that destroyed the school's main building. According to a story which Cassius did not divulge for nearly seventy years, the fire was the fault of a body servant whom he kept at school with him. About midnight, on Saturday, May 9, 1829, Clay was asleep in his room while the slave boy polished boots out on the stairs. To illuminate his work the boy placed a candle upon the top step. Before he finished, however, he fell asleep and did not wake until the candle had burned down and set fire to the stairs. "The flames went like powder," recalled Cassius, adding that he "ran down with some clothes in my hand in my night shirt." The two scared boys barely escaped from the burning building. The imposing structure was soon a smouldering shell of ashes. Except for a few books, the building and its contents were a total loss. Cassius Clay's sleepy slave had caused the destruction of a valuable piece of property, and the experience awed Cassius as few things did. He kept his secret well through the years. The college, already weakened by a religious controversy of long standing, never overcame the effects of the fire. By the time a new building replaced the old, other colleges—many of them denominational—had appeared, and Transylvania lost its favored position. Cassius Clay, through his slave boy, had unwittingly precipitated a crisis in the affairs of the college. It was not the last time that he would be involved in changes reaching far beyond his own life, even into the course of state and national history.[20]

During his stay at Transylvania he encountered, under less destructive circumstances, another influence that would help to shape his life. He met Mary Jane Warfield, the girl he was to marry a few years hence. In 1829 she was a sixteen-year-old pupil in a Lexington school, with a fair complexion, limpid, gray-blue eyes, and light-brown hair which flowed down her back in long silken curls. The impulsive Cassius was immediately attracted to her. She was the second daughter of Dr. Elisha Warfield, a Lexington physician better known for his racing stable than for his medical skill. Her mother was Maria Barr of Lexington. The girl's parents Cassius would learn to dislike, and with Mary Jane herself he would quarrel seriously, but all those troubles lay in the future. As a student, filled with youthful fervor, he saw only her appealing beauty—she was the "impersonation of 'eternal springtime.'" He composed poems to her, and before he left Lexington, he bade her a sentimental and scholarly farewell. To Mary Jane, Cassius presented the prize book—Washington Irving's *Sketchbook*—he had been awarded at college. Inscribing a few lines of Byron on the flyleaf, Clay paid the young girl a graceful compliment. She would not forget so polished a courtship.[21]

Bidding farewell to Mary Jane and to Transylvania University, Cassius embarked upon an extensive journey, "for the purpose of observation and improvement," as he explained it. Carrying letters of introduction to President Andrew Jackson and to other political notables, he visited Washington and other eastern cities. He professed a desire to learn, but he was not interested in historical sites or museums. "I shall attempt to make it my business whilst here to visit and examine men rather than their buildings and inanimate curiosities," he said of his journey. Already the desire to study men of power was growing within him; he wanted to meet successful politicians so that he might discover the reasons for their success.[22]

With that object in mind he sought out the leading political figures of New England. He was introduced to aging former President John Quincy Adams, and to Senator Daniel Webster, then basking in the glow of the publicity that followed his debate with South Carolina's Robert Y. Hayne. In Boston, Clay became acquainted with many people later connected with the antislavery movement: the poet John Greenleaf Whittier, the song-writer Julia Ward Howe, the lawyer John A. Andrew, who became Civil War governor of Massachusetts, and the eloquent orator Edward Everett. Contact with such people and exposure to their ideas made Clay's journey through the East an experience which added other facets to his maturing personality.[23]

Clay had come North not only to study politicians but to continue his formal education. In 1831, he enrolled in the Junior Class at Yale College. His experiences in New Haven had a lasting influence upon Cassius Clay, and what he learned there completed his training.

At Yale he met people he never forgot. Elderly Jeremiah Day, author of mathematics textbooks, was the college president, and Professor Benjamin Silliman, "of large stature and of large brain," as Cassius described him, was an innovator in chemistry and natural science courses in American colleges. James L. Kingsley was Clay's instructor in the classical languages, and Denison Olmstead was professor of mathematics.

Among his classmates, two in particular remained in his memory: Allen Taylor Caperton, of Virginia, the class practical joker who became serious enough in later years to win a seat in the Confederate Senate; and Joseph Longworth, later known for his philanthropies to the city of Cincinnati.[24]

Clay fitted easily into the New England academic community. The Yale class he entered was small; it numbered only fifty-seven members because in the previous year, as a result of the notorious Conic Sections Rebellion, many men had been suspended or expelled. The course of study Cassius followed at Yale was similar to that at Transylvania. He studied the traditional Latin and Greek and also philosophy, rhetoric, and history. He continued his interest in debate and oratory and joined a Yale literary society. He made a name for himself as a speaker, and in scholarship he stood high in his class. In the examinations of September, 1831, only two students had higher marks than he.[25]

But he was not too studious to participate in college pranks. "I've . . . been on the point of going to gaol," he confided to Brutus. To finance the fun was not easy, and he complained that New Haven was a hard place to live in, "hard to get money, hard to keep it, and still harder to do without it." Many times he repeated the student's prayer: "Preserve me from anything but a full pocket!"

Clay took part in the often ribald immaturity which accompanied student life, but he was beginning to get bored with college. To a young gentleman in his early twenties it had begun to appear childish, and he longed to begin a man's activities. He had been at school, he moaned, long enough to make any man "an artificial if not a natural fool."[26]

Despite his increasing distaste for study, Cassius did not relish the thought of returning to Kentucky as a planter. His contact with men in high places, and the new ideas he met in New Haven, made him averse to a life on the plantation. "It is with dread that I think of plough and hoe," he confessed. And before he left Yale he had determined upon a political career which would make his farm merely an interesting avocation.[27]

Not all Cassius' new ideas came from the classroom. A chance encounter with William Lloyd Garrison, who had recently begun the publication of his uncompromisingly emancipationist journal, the *Liberator*, made a lasting impression upon him. Cassius heard the impassioned editor speak at New Haven's South Church, and for the first time in his life he heard a straightforward denunciation of slavery. The year before Clay came to New England, he had joined a Kentucky emancipation society, but that action was no measure of his opinion upon the subject. He regarded slavery as so many other southerners did—the "fixed law of Nature"; he thought little about the institution and when asked about it was apt to say that nothing could be done about it. But Garrison, fiery and provocative, suggested a drastic solution: immediate abolition of slavery and no union with slaveholders. Though Cassius would reject Garrison's extreme program, the idea of doing something to end slave labor became a significant part of his creed. It was his part in the antislavery crusade that brought him to the attention of Abraham Lincoln and that eventually won him an appointment as Minister to Russia.[28]

In New Haven Clay also observed the effects of free labor upon the economy, and it was the conclusion he drew from his observations that most profoundly influenced his future action. In New England he was surprised to see an industrious people reaping prosperity from an unfriendly soil. In the South, where soil fertility was almost the only criterion for judging economic standards, these people would have been among the poorest, but here industrialism enriched the population. "I . . . saw a people living *there* luxuriously on a soil which *here* would have been deemed the high road to famine and the almshouse," he reported later, back in Kentucky. Connecticut, which he had been taught was a land of "wooden nutmegs and leather pumpkin seed," was in reality a land of sterility without paupers. Seeking an explanation for the discrepancy he saw between New England and his native state, he concluded that "*liberty*, *religion*, and *education* were the causes of all these things."

Cassius Clay had fitted the final plank into his life's platform. For the South, he wanted economic prosperity; for himself, political success. He blamed slavery for the economic and social inferiority which existed in Kentucky and in the South, and he based his opinion upon his personal observations while a student at Yale. "Nothing but slavery," he maintained, was the cause of the economic backwardness of the agricultural South. That motto became the central theme of his career, and he never went far beyond it; indeed, in years to come it became a restrictive monomania which stood between him and his objectives. But reiterate it he did. Whenever he had to make a decision, the foundation for his choice was the antislavery argument he had developed during these early years. It made him a revolutionary in the 1840's and 1850's, but before his public career was over it would make him a pathetic anachronism.[29]

It was Clay's fidelity to this principle throughout his lifelong pursuit of public office that made him unique. Many another southern student had made similar remarks about slavery while on a northern campus and then had forgotten them when he returned home. Not so Cassius Clay. The more he pondered the basis for a successful political career, the more he became convinced that opposition to slavery was the proper course. He had gone North to study the ways of successful politicians, and it was his luck to stumble upon the issue which would dominate political affairs in America for the next generation.

Even in 1832 he recognized its political implications. "The slave question," he wrote Brutus from New Haven, "is now assuming an importance in the opinions of the enlightened and humane, which prejudice and interest cannot long withstand." The slaves of all the South "must soon be free," he predicted, or there would be a dissolution of the Union within fifty years, "however much it may be deprecated and laughed at now. . . ."

Here was a potent issue; it appeared to be growing in significance; and the prognostications were that slavery was doomed. To the young but ambitious Clay that was enough. For the remainder of his life, perhaps as much from stubbornness as from principle, he would plod the antislavery path, using the same reasoning he had employed as a student at Yale.[30]

Clay's first public declaration of his new faith came as an

idealistic afterthought to a florid oration. His abilities as a public speaker had attracted the attention of his classmates, and on February 22, 1832, the Senior Class chose him to deliver an address commemorating the centennial of Washington's birth. Eulogizing the Father of his Country, Cassius concluded with a peroration. He described the day as one of rejoicing over national liberty, the gift of Washington, and asked, "Does no painful reflection rush across the unquiet conscience?" Picturing Washington's admirers bringing gifts to freedom's altar, he challenged, "Are there none afar off, cast down and sorrowful, who dare not approach the common altar; who cannot put their hands to their hearts and say 'Oh, Washington, what art thou to us? Are we not also freemen?' " In that indirect reference Clay mentioned the slaves in the United States, and upon his masked reference he built an emancipationist injunction. "Foolish man," he said, "lay down thy offering, go thy way, become reconciled to thy brother, and then come and offer thy offering." The subtle language, drawn from the Bible, hinted at his growing opposition to slavery. In years to come he would make his meaning entirely clear, and he would drop the religious phraseology.[31]

Clay used the Washington's Birthday oration to express publicly his fear that the Union might some day dissolve. "It needs not the eye of divination to see that differences of interest will naturally arise in this vast extent of territory," he said. "Washington saw it; we see it." In the political arena, he pointed out, alluding to the Webster-Hayne debate, "the glove is already thrown down; the northern and southern champions stand in sullen defiance." He hoped, Clay added, that a "blinded people" would not "rest secure in disbelief and derision, till the birthright left us by our Washington is lost! till we shall be aroused by the rushing ruins of a once 'glorious union.' "[32]

The oration might well have been Clay's graduation address.

Though he did not graduate for several months, the speech revealed that his training was complete. Of the many factors which had gone into his personality—the influence of his parents, his environment, his schooling in Kentucky and in Connecticut—the most significant had been his sojourn among the industrious New Englanders.

As he left New England and headed homeward, he was no longer an adolescent scholar but a mature individual starting forth upon a career. Despite his youthful appearance, he was an adult. He looked young, innocent—almost feminine. His face was fair and soft, his eyes dark and thoughtful, and his hair, shaped by a stubborn cowlick on the left side, was long, black, and silky. Though he displayed none of the fire which characterized his later life and made him a legendary figure in Kentucky history and the subject of international gossip during his stay in Russia, there was in his eyes a bold look of determination.

In stature he had completed his growth; he was well over six feet tall; but his shoulders were not so broad as they would become in another decade. Endowed with a magnificent physique, an aggressive and courageous personality, and a driving ambition, Cassius M. Clay, combining character and training, would make an exciting bid for political power in his native state. His efforts there, as an antislavery politician and a pioneer in the movement that culminated in the establishment of the national Republican Party of the 1850's, were to bring him a country-wide reputation as the Lion of White Hall. It was the part he played in Kentucky politics that determined the course of his subsequent career.

CHAPTER II

A CONGENIAL CAREER BEGINS

WHEN Cassius Clay returned home from study at Yale College, he brought with him more than the newly embossed diploma in his baggage. In his opinions he was not the same person who had left Kentucky in 1831. Having lived for over a year in busy New England, he had come to admire its prosperous economy. He had pondered the problem of Kentucky's inferiority in the external trappings of prosperity: railroads, factories, and schools. In the future, as he surveyed his home state with the analytical eye of a political theorist, he would not forget the energetic Yankees who outwitted miserly Nature by mechanical devices. The vivid memory of New England's industry continued to influence the course of the career he was about to begin.

As Cassius had altered his opinions while he had been away, so the state to which he returned had also changed. The momentous issue of nullification had appeared in the interim, bringing to the surface again the quarrel between industrialists and agrarians for control of national affairs. South Carolinians, faced with the dilemma of decreasing soil fertility and diminishing returns from their cotton and rice fields, blamed their difficulties upon a convenient scapegoat, the tariff. It made sense to them: if they could buy British goods at lower prices, it stood to reason that they would live better. But

Yankee manufacturers preached economic nationalism, urged Americans to buy at home, and sent their representatives to Congress with orders to demand higher tariff protection. Gentlemen of the South Carolina plantations resisted the Tariff of 1832 with the nullification argument, ably presented by their political knight-errant, bushy-browed John C. Calhoun. The states were the guardians of the Constitution, they argued; any unconstitutional action of the national Congress should be nullified by the states until the Constitution was amended or the proposed law overthrown. President Andrew Jackson met this theoretical attack upon national power as he had met the southern Indians: he threatened force, and the South Carolinians had to accept a compromise. The canker on the body politic had thus been bandaged, but the sore remained. Neither side had been satisfied with the settlement; New Englanders disliked the reduced tariff rates, and southern spokesmen protested the obvious threat to their interpretation of the Constitution. As Cassius Clay had said in his Washington's Birthday oration, the protagonists "stood in sullen defiance." The tug-of-war on the national scene between industrialists and agrarians was to be re-enacted in Kentucky, where Clay would play a significant role in the drama.

In Kentucky the fight involved the American System, the brainchild of the state's favorite son, Henry Clay. A clever attempt to synthesize interest groups, the American System was a scheme to win support of New Englanders and western farmers by offering public assistance to industry and to food-producers. To encourage the establishment of home manufactures in the Ohio valley as well as in the East, Henry Clay advocated higher and higher tariff measures. He championed internal improvements at public expense, to enable western farmers to market their surplus, as well as to aid western commercial development. For business interests all over the country, the Gallant Harry favored the National Bank, which would provide a firm monetary and credit system. His was a plan which anticipated a self-sufficient economy within the country and bound the sections together with a golden chain of commerce.

Such was the platform upon which Henry Clay had for years sought to unite the divergent interests of the nineteenth-century Federalists, who would soon take the name Whig. Opposition to the scheme appeared in many areas of the South, particularly among slave-owners. They could discover little benefit to themselves from the promises of Henry Clay's system, and they considered any tariff a tax which fell unequally upon them. Trusting in the soil and in the philosophy that "Cotton is King," they were not interested in the economy of the mill and the machine. Many planters who objected to the Whig Party because of its industry-centered American System spoke out against it; the elder Robert Wickliffe, the "Old Duke," Kentucky's most prominent slave-owning planter, remarked a few years later, "For myself, I am opposed to Mr. [Henry] Clay, because I consider him a dangerous politician to the whole of the American States, and especially to the Southern and planting States." Cassius M. Clay, born to the class represented by Wickliffe but fresh from New England, where he had seen the potentialities of the American System, was familiar with the essential elements of both conflicting attitudes toward economic prosperity.[1]

Because he returned to Kentucky and began his political career at the time when party alignments were shifting, Cassius was forced to choose between the agricultural economy of the planter-aristocracy to which he had been born and the vision of industrial prosperity which he had brought from New England. To be successful as a politician, he would have either to follow the dominant slaveholding minority—which opposed the Henry Clay program—or to espouse the cause of Kentucky's embryonic proletariat, which had as yet no political consciousness. Neat compromises were impossible: the immoderate invective of William Lloyd Garrison and the bloody slave-insurrection in Virginia made that clear. At the beginning of his career, Cassius Clay was not fully aware of the disagreements between the planter and the white artisans, and he attempted to serve the interests of the laborers while remaining a member of the planter class, as southern politicians had done before.

But Cassius' political experiences were proof that the party of Henry Clay faced a fundamental conflict in the South. Henry himself maintained his position in Kentucky only because of his astute abilities as a compromiser; Cassius, when he took an extreme stand in support of the same views Henry advocated, ran afoul of local prejudices. But Cassius remained stubbornly loyal to the conclusions of his New Haven years. He advocated a southern industrialism and became a spokesman for southern non-slave labor. Throughout his career he consistently defended the economic principles of the American System. When he entered Kentucky politics, these views were still acceptable, but as time passed and sectional loyalties fermented, the economic principles of Henry Clay became more unpopular, and Cassius was persecuted for failing to uphold the glories of agrarianism. Men who in 1835 were involved with him in commercial ventures were, ten years later, castigating him as a traitor to his state.

The view which he would not relinquish, and which became the theme of his entire career, was an interpretation of the idea of his kinsman Henry Clay to suit the southern situation. As the American System envisioned a balanced national economy, so Cassius sought to increase Kentucky's prosperity by broadening its economic base. He wanted to apply Henry's program of public aid to commerce and industry in his state in order to encourage a diversified, balanced economy. A border state, Kentucky enjoyed the distinguishing features of the planting South and the manufacturing North; rich farmland and a plantation economy existed beside a mountainous, mineral-rich area potentially analogous to industrial

New England. Cassius wanted to encourage his fellow Kentuckians to capitalize upon their favorable situation and erect a manufacturing economy in the southern mountains, or "American Switzerland," as he called them. To meet the needs of the industrial workers whom he hoped to entice to the state, farmers, he urged, should emulate their counterparts north of the Ohio by growing food-grains and meat rather than slave-produced staple crops.

Such a course would increase the state's income, he promised, and would enrich the free white mechanics—the "middle class," as he termed them. Moreover, he contended that farmers would also prosper by the addition of a home market for foodstuffs, which was lacking under the predominant plantation system. Cassius Clay's plan paralleled the American System on the state level and had a similar purpose—to unite the sections and to elevate a party to power. It was, therefore, a Kentucky System that he proposed in his bid for political office. But the great enemy to the success of the scheme was the planting class, proponent of an agrarian economy, which was also the proslavery party. Because they sought to obstruct his plan, Cassius would condemn the planters as opponents of progress and as aristocratic oligarchs heedless of the majority interest.[2]

Many years were to elapse, however, before Cassius Clay would complete the details of the Kentucky System. When he did, it was the tragedy of his life that his belligerent personality and the prejudices of his fellow southerners deprived him of his reward.

The early years of his career Cassius spent in an effort to establish himself in his community and in the sight of its political leaders as an available candidate for responsible office. His first step in that direction was to get a background in law. After a summer's vacation from study, he entered Transylvania University for the second time. He was happy to return to the pleasant community of Lexington and to escape the loneliness of White Hall. His mother had remarried while he was away and had moved to Frankfort; all of the children had established separate family homes, and the big house was bare and quiet. For a time he busied himself reading law, and concluded a six months' course, but did not take out a license to practice. His purpose in studying law was not to become a lawyer, he explained, but "to prepare myself for political life, which was congenial to my taste." [3]

And in addition to its law school Lexington held other attractions for Cassius. Mary Jane Warfield, now two years older and in the full bloom of youth, was more charming than ever, and there were competitors for her favors. She preferred the dark-haired, dashing Clay, however, and he was certain of his love for her. Though members of his family warned him away from the Warfields, Cassius impetuously overrode their objections and determined to win her hand. In asking Mary Jane's father for permission to marry her, however, Cassius made his first mistake. He soon discovered that his troubles were just beginning. Dr. Elisha Warfield was a small, shy man, who allowed his domineering wife to run his household while he contented himself with running his race horses. Because Green Clay had been undisputed master at White Hall, Cassius did not understand the situation in the Warfield family. Instead of presenting his request to the matriarch, the unsuspecting youth aroused a miniature tempest by dealing with the father. After he had smoothed that difficulty, Cassius imagined that all was ready for the wedding. Mary Jane shared his assurance and set February 26, 1833, as the date for the ceremony.

But the arrangement of the marriage was not to be that easy. Mary Jane's mother, still irritated by the whole affair, showed Cassius a letter she had received from Dr. John P. Declary, a Louisville physician and one of Mary Jane's rejected suitors. Declary told Mrs. Warfield that her daughter should not marry a Clay, and in particular she should avoid the young, unsettled Cassius. When Clay saw Declary's charges he felt compelled to vindicate himself in the eyes of his bride-to-be. With his best man, James S. Rollins, as his second, Cassius journeyed to Louisville to obtain a public apology.[4]

Armed with a stout stick and the derogatory letter, Cassius and Rollins invaded Declary's hotel and invited that gentleman outside. With his stick under his left arm, Cassius showed the doctor the passages in the letter to which he objected and offered him the opportunity to explain them. "What do you have to say?" Clay asked him. Though the doctor was ten years older than Clay, the two men were about the same size. When the doctor made ready to fight, Clay wrestled with him in the street, and then brought his stick into play. Repeatedly he belabored the unarmed man each time he struggled up from the ground. When passers-by ran forward to intervene, aghast at the beating Clay was giving the doctor, Rollins held them off with a gun. When the doctor no longer tried to rise, Clay told him where he was staying, and, obedient to the code, returned to his room to await the challenge. Within a few hours it arrived; Declary would meet Clay across the river in Indiana the next day, Sunday, February 25, the eve of Clay's wedding.

Early the next morning Clay and Rollins crossed the Ohio in rented carriages, and Clay practiced firing his pistol at a tree while awaiting his adversary. But by the time Declary and his party arrived, a host of curiosity-seekers had gathered upon the scene to witness the anticipated blood-letting. The contestants decided to postpone their fight and meet later back in Kentucky. Through a series of comic misunderstandings, for which each side blamed the other, the duel never occurred. As night fell, Cassius was more concerned about getting back to his wedding than he was in salving his wounded pride. After waiting all night for an attack, Clay and Rollins caught the morning stage to Lexington and considered the challenge off.[5]

Late Monday night, February 26, Clay arrived, mud-spattered and travel-weary, at The Meadows, Mary Jane's home near Lexington. His bride-to-be had waited there all day, not knowing whether he was alive or not. In her life with him she would endure many other days and nights of loneliness in the same uncertainty. But now she donned gown and veil and took her vows. At The Meadows, in the candle-lit main parlor, Cassius M. Clay and Mary Jane Warfield were married by the Reverend B. O. Peers, an Episcopal minister, president of Transylvania University.[6]

But the happy conclusion of Clay's courtship did not end the threat to his marriage. In a series of public letters, Declary belittled Clay's courage, charging that he had fled from the Louisville challenge. At first Clay tried to ignore Declary's insults. "For a man to leave a newly-married wife to return to fight her rejected suitor," Clay remarked, "was too absurd for even the fool-code." But the taunts of his friends soon got under his skin, and he determined to "give Declary a full test of his manhood." Taking a long pleasure trip to Cincinnati and St. Louis, Clay returned to Louisville. Arming himself, he went to Declary's hotel, the Old Inn, and sat in the lobby awaiting the arrival of the doctor. When Declary saw Clay he turned pale and walked away. Cassius remained in the city for several days, but since his enemy made no contact with him, he returned to Lexington. The next day he heard that John P. Declary had gone into his room, locked the door, and with a razor slit his wrist arteries and bled to death. He was, a Louisville editor commented, killed by a duel that he never fought. Not all of Cassius Clay's enemies would be vanquished so easily.[7]

Having provided White Hall with a suitable mistress, and having completed his legal training, in 1834 Cassius announced his candidacy for the lower house of the state legislature. In listing the contenders for the office, a Whig editor declared that Clay was "*of the true faith.*" But despite the recommendation, Cassius had to withdraw from the race. His opponents pointed out that the state constitution required that a legislator must have passed his twenty-fifth birthday before the opening of the session, and Clay would not be that old. "Upon a more critical examination of the constitution," explained the twenty-three-year-old Clay, somewhat embarrassed, "I am convinced that I am not of lawful age." Claiming that his private affairs required his attention, the impatient politician announced that he would wait a year before offering himself as a candidate.[8]

The year was a busy one for Cassius, who used it to enhance his availability as a Henry Clay Whig of the American System. In addition to his own affairs, which took him all over the state, he participated in public-development enterprises. It was a year in which internal improvements boomed, and Kentucky was no exception. When a company organized to construct a turnpike road from Lexington to Richmond, Kentucky, seat of Clay's home county, he took an active part in its plans. If he could influence the road survey, its route would traverse much of his land, and would cross the Kentucky River at his ferry. Cassius served as secretary to the road company—sitting with men who later denounced his course—and he was also a commissioner in the Richmond branch of the Northern Bank of Kentucky.[9]

Clay's youthful devotion to business principles revealed his understanding of the Kentucky political scene. In the 1830's, the basic issue that divided parties was the place of government in the affairs of men, particularly economic affairs. Generally, the groups which were coalescing under the name Whig argued that government should use its financial and legal power to improve business conditions by construction of roads and canals, by assistance to railroads and river traffic, and by establishment of a banking system and a sound currency. "We indulge the hope," a Whig editor remarked, "that the time is not distant when our state will be enabled to vie with any other in the Union in those great works of enterprize, which must redound so much to the wealth, prosperity, and comfort of her citizens."

The Democrats, on the other hand, favored frugal government with low taxes and restricted powers. They opposed public assistance to internal improvements, calling it unconstitutional. President Andrew Jackson had brought the matter to a head by vetoing the Maysville road bill, a project in which the national government was to take stock in a company organized to construct a road from Maysville, Kentucky, on the Ohio River, to Lexington. "If it be the wish of the people that the construction of roads and canals should be conducted by the Federal Government," Jackson said, "it is not only highly expedient, but indispensably necessary, that a previous amendment of the Constitution . . . be made." Jackson's veto message provided Kentucky's Henry Clay with the issue he needed to arouse resistance to the President and to his party. But in the meantime, Kentucky businessmen, having failed to receive assistance from the national government, looked to Frankfort for aid. In the state, accordingly, it was the Whigs who represented business and commercial interests, and these Cassius joined.

But in the mid-1830's another factor complicated the Kentucky political alignment. President Jackson's militant action against South Carolina in the nullification controversy, as well as his efforts to win support of New England workers, had aroused resentment and derision among southern landowners. That group made first use of the name Whig as a motto for their opposition to "King Andrew the First." Cassius was not a "Cotton Whig," as they came to be called, but he had been nurtured to respect the landholding aristocracy. When he began his career, therefore, his business as well as his family connections made him acceptable to Whig leaders.[10]

His acceptability soon brought its rewards. In the summer of 1835—as soon as he reached the requisite age, he loved to point out—he was elected to the Kentucky legislature as representative from Madison County. As a member of the General Assembly, Cassius continued his support of the American System. Already active in public works, he established himself as a proponent of government assistance to internal improvements. He worked to obtain a state subsidy to finance a railroad from Louisville to Charleston, South Carolina, which would cross his county. He sponsored a petition from the president and officials of the Lexington and Richmond Turnpike Road Company, of which he was an officer and a stockholder, and he utilized every opportunity to express his approval of protective tariffs.[11]

Clay also expressed his distaste for Negro slavery as a labor system but refused to agitate for its abolition. For some years there had been in Kentucky a strong current of opinion, especially among those who favored gradual emancipation, for amendments to the slave clauses in the state constitution. Such a course required a constitutional convention. In debate upon a bill to authorize one, Cassius Clay protested that it was not the proper time to discuss the slavery issue. "Is this a time, when a horde of fanatical incendiaries is springing up in the North, threatening to spread fire and blood through our once secure and happy homes," he asked, "to deliberately dispose of a question which involves the political rights of master and slave?" A convention, he said, might provide for an emancipation program which, on the surface at least, appeared wise. There had been, he confessed to his colleagues, a time when he favored gradual emancipation, and he still saw advantages in free labor. Having compared slave and free communities, he said, "I am candid in saying that the free states have largely the advantage." As he spoke, a vision of New England returned to him. "I cannot, as a statesman, shut my eyes to the industry, ingenuity, numbers, and wealth which are displaying themselves in adjoining states." But, he continued, the "spirit of dictation and interference . . . in the North" precluded any chance for improvement. "I almost give way to the belief that slavery must continue to exist till, like some ineradicable disease, it disappears with the body that gave it being." Clay was not yet ready to advocate a cure, but his simile was clear: slavery was a grim plague which would destroy the society that harbored it. Like other Kentucky gentlemen, however, he maintained a passive attitude toward it. Although he opposed calling a constitutional convention, a device he would advocate a few years hence, his unfavorable description of slave labor and his admiration for an industrial economy foreshadowed the conflict which lay ahead for him.[12]

After the assembly had adjourned and Clay had returned home, he took an increasingly active part in Whig affairs. On April 19, 1836, he attended a party convention in Lexington and served on the resolutions committee, and in the summer, at the Young Men's Whig Convention in Louisville, he became a member of the influential committee of correspondence. Despite his advancement in the state party, in his home county his campaign for re-election was swallowed up by the Van Buren sweep of that year. On August 5, he told his brother, "The election is over, I am beaten," and blamed his advocacy of internal improvements. John F. Busby, an adversary of public assistance to roads and canals, replaced Cassius at Frankfort. Many years later, Clay explained his defeat in the homely language of the tobacco-grower, saying that his friends decided to "top me, and let me spread." [13]

Ousted from the assembly, Cassius retired and again turned his attention to private affairs. In the face of personal tragedy —his first two sons died in infancy—Cassius worked at the routine business connected with his father's estate, and invested in new commercial ventures. He formed a partnership with George Weddle, of Madison County, in a sawmill on the Kentucky River and established a gristmill near that stream. He also continued to operate the river ferry which his father had begun. While he did not invest in heavy industry, still he had economic interests other than hemp or tobacco, the customary produce of Kentucky plantations. He specialized in beef cattle, which he drove to the Cincinnati market, and he became a nationally recognized authority on the breeding of Shorthorn cattle. He also offered purebred Southdown sheep and Spanish hogs on a national market. Thus, even before the panic of 1837, Cassius Clay had enlarged the scope of his financial operations beyond the plantation. When his political views looked beyond the confines of a narrowly specialized agrarian economy, they merely reflected the realities of his own sources of income.[14]

It was out of his personal economic situation that Cassius Clay's political philosophy emerged. His experience was not at all unique, but it led him into an unpopular position, and there his stubborn courage kept him. In the spring of 1837, he entered the legislative race, again as a Henry Clay Whig, and this time won easily, finishing first in a four-man race for the two Madison County seats. But during the session Clay began to manifest those differences which would soon set him apart from his neighbors.

He used every opportunity to serve commercial interests, particularly along the Kentucky River route to the mineral-rich mountains. He soon charged that slaveholding agriculturalists were his chief opponents in that program. Cheap-money men in the General Assembly, like Robert Wickliffe, Sr., of Fayette County and William R. Evans of Monroe, advocated a bill conferring banking privileges upon the proposed Charleston and Ohio Railroad. Cassius Clay, committed to commercial development within the state but opposed to cheap money, took issue with them and associated the pro-slavery party with those who opposed the American System. "There is a class of politicians who have solemnly declared themselves at war with the system of American manufactures," he announced. "There are men who have avowed themselves inimical to a system of internal improvements." The same group of men had succeeded in destroying the "best bank circulation among any people," and now they desired to flood the state with new issues of paper money. They also demanded repeal of the tariff, to import "at a sacrifice, from foreign and alien merchants, kingly subjects, rather than sustain the freemen of our common country." Clay became even more indignant; it was the same group, he charged, which kept alive the slave question, "that question which of all others is most terrible to the hopes of this union." [15]

Cassius, stubbornly defiant and determined, was not a man to back away from a fight, either political or physical. He engaged in several spats upon the floor of the House, and it was a brawl with a colleague that illustrated once more his penchant for adopting a driving attack. James C. Sprigg, representative from Shelby County, had once made the mistake, while drunk, of confiding to Clay his dearest secret: when he had a quarrel with anyone, he had boasted, he would await no formal preliminaries, but would strike without warning and thus gain the victory. Clay tucked the bit of information away in his memory, for that was also his technique, as one George, slave boy at White Hall, could attest. As it turned out, Clay made good use of his intelligence concerning Sprigg's battle plans.

In the House Sprigg and Clay had frequent tiffs, for the most part over the work of the Banking Committee. On several occasions, Sprigg made bitter remarks about the alleged inefficiency of the members of the committee. The other members of that group appeared to ignore him, but the excitable Clay regarded his words as personal insults. Finally, after an unusually heated speech of Sprigg's, Clay rose to respond.

After the House had adjourned for the day, Clay met Sprigg in the hotel. Sprigg walked up to him, wearing a pleasant look, but Cassius was aware of his colleague's tactics. When Sprigg came within reach, Clay suddenly struck him a severe blow with his huge fist, full in the face, and sent him staggering to the floor. Holding the advantage, Cassius clubbed him again and again with his bare hands until bystanders rescued the badly beaten Sprigg from the vicious hammering. Sprigg later confessed to Clay that his intention had been to strike without warning, and he could not understand how Clay had divined his purpose. To Clay, the incident proved that the best defense was a strong right arm and the determination to use it.[16]

But he was also endowed with a fine voice and a clear mind, as well as physical strength, and these qualities attracted attention to his abilities as a legislator. "Mr. Clay, of Madison, spoke with much force and wit," an assembly reporter said of him. "This gentleman possesses a fine flow of words and ideas; and with proper application he must make a cynosure in the commonwealth."[17]

But trouble, not success, lay ahead for Cassius Clay, for in his ambitious plans there was the seed of conflict. In his own financial program he had moved away from the plantation economy, and he stood to lose if industrial expansion was halted. Yet the dominant element in Kentucky, the very group which might provide him with the career he craved, stood hostile, though not united, in regard to that program. Nevertheless, he would not change his views. His own economic stake and his deep convictions about the basis of public welfare, as well as his native determination, held him loyal to the creed he had adopted. His loyalty would meet severe tests.

The issue did not come to a head for two years, however, and in that time Clay experienced both humiliation and honor. First, he prepared the way for the major political battle of his early career by moving his legal residence to Lexington. To find a more central place for his political aspirations, he moved his family to the seat of Fayette County, rich in tobacco, hemp, race horses, and bourbon whisky. The county was also the center of Kentucky's slave-owning plantations, with over ten thousand Negroes. Fayette offered greater opportunity for political prestige, and there was another reason for the move: Mary Jane was bored with farm life and longed to live closer to her parents and friends. So, for political and for personal reasons, Cassius decided to change his headquarters. In Lexington he purchased Thorn Hill, an elegant and comfortable residence in the city.

Clay's acquisition of Thorn Hill belied the embarrassing condition of his personal finances. In the panic of 1837, he had committed himself heavily in an effort to pay the obligations of William Rodes, husband of his sister Paulina. Much of Green Clay's estate was swallowed up in the crisis. Clay also had troubles of his own; his business partner, George Weddle, failed in his attempt to manage the sawmill and ferry. In a day of unlimited copartnership liability, Clay suddenly found himself besieged by creditors and unable to pay them. Indeed, he could not settle his debts for twenty-five years. His indebtedness affected his later actions; his lecture tours and his journalistic enterprise were, in part, attempts to meet his obligations.

Though Clay's financial situation was unhappy, his political career was bright. Until he became a legal resident of Fayette County he could not seek local office, but he attracted the attention of state Whig leaders. On December 4, 1839, he sat as delegate to the Whig national convention at Harrisburg, Pennsylvania, and served as a floor leader for the candidacy of

Henry Clay. Despite Clay's efforts on behalf of his kinsman, the nomination went to the hero of Tippecanoe, William Henry Harrison. The result, Clay later recalled, was a bitter disappointment for him. He wept, "overcome with a sense of injustice" that his candidate had been "betrayed." [18]

CHAPTER III

CLAY TAKES HIS STAND

IN THE 1840 race for a seat in the state legislature from Fayette County, Clay became an active critic of the slave system and of its defenders. For that reason the campaign was the first overt step in his growing rebellion against the prejudices of his class. But in addition, since his proslavery competitors were also Whigs, the election accentuated the divisions within the party.

The chief issue upon which the Kentucky Whig Party split, as illustrated by Cassius Clay's campaign, was whether the economic interests of the state should emphasize exploitation of land or should expand to include mechanical skills. That question was soon overshadowed by another, more emotional, dispute between partisans of slave labor and proponents of free white labor. Agrarians rallied to the support of slavery, prevalent on the plantations, while advocates of the mill posed as humanitarian liberals interested in freedom and justice. Within the party the split was between Cotton Whigs and Conscience Whigs.

The conflict became a political issue when Cassius Clay announced himself a candidate for the legislature. At that time Fayette County returned three members, and there were already three candidates in the field, all Whigs. John Curd and Clayton Curle were incumbents and were practically assured of re-election. The third contender, whom Clay challenged, was Robert Wickliffe, Jr., who was dominated by his powerful father, the Old Duke. Clay and the younger Wickliffe had much in common: they were about the same age; they had been classmates at Transylvania; and they were both rising career politicians with legislative experience. There was, however, an important difference between them: Wickliffe was a native of Fayette County, but Cassius was a newcomer.

To have any chance at all for victory over Wickliffe, Cassius needed an issue which would make him appear a better supporter of Fayette County interests than the hometown candidate. He found it in the so-called Negro Law of 1833, a prohibition against further importation of slaves into the state. The restriction upon the interstate slave trade limited the growth of the Negro population, thereby acting as a tariff to benefit the Bluegrass, an area already well supplied with slaves. Any planter or enterpriser who required more Negro labor had to procure it within the state; a scarcity of such labor accordingly enhanced the value of Bluegrass slave property. "Slavery being unequally diffused," argued one legislator, "the law is a bonus to Fayette, Clark, and Bourbon [counties in the Bluegrass], where they own large numbers of slaves." The import restriction enabled "these rich counties to sell, to constituents of other areas, slaves which are now commanding a price, sometimes reaching . . . the extravagant sum of $1400 each." [1]

Because of their monopolistic supply situation, Kentuckians in counties with large slave populations might be expected to favor the law. But there was another aspect to the problem. The import restriction was regarded as a means of gradually ending slavery by reducing the proportion of Negroes in the total population. "There are thousands upon thousands of the citizens, and themselves slaveholders, too," commented the editor of the *Frankfort Commonwealth*, "who look upon slavery as an evil. . . . They have considered that an act which operates to keep down the increase by excluding foreign supply [would aid the] numerical growth of the white race." In the course of twenty or thirty years, the editor prophesied, the percentage of blacks would be so small that slavery would scarcely be felt in the state. There may have been many Kentuckians who wished to restrict the growth of slavery in that manner, but there were also Kentucky slave-owners who had no intention of losing either their slave property or their dominant political position without a struggle. If the Law of 1833 would reduce slavery to an insignificant position, they demanded an end to the import restriction even if it meant an end to their monopoly on the supply. Such a slave-owner was Robert Wickliffe the elder, of Lexington, who spoke for his compliant son. Referring to the Negro Law as an "abolition tinder-box," the Old Duke denounced those who sustained the import restriction. "That Kentucky is to be the first battleground of the abolitionists, they all agree," he said. It was the antislavery object to drain the slave population from the state, he declared, and to "shut out emigrants from slave states, until the price of slave labor shall rise so high that a poor man cannot command it." The law was an "implement in the hands of the abolitionists, to carry out their views in regard to our slave property." For that reason, Wickliffe urged the repeal of the measure.[2]

The double face of the Law of 1833 made it an admirable issue for Clay to use in his campaign. In a county with nearly ten thousand slaves, the politician who advocated repeal of the restriction—albeit in order to protect the institution of slavery—would lay himself open to the charge of neglecting local interests. Discussion of the Negro Law pushed the Wickliffes to that position, for the question soon entered the campaign. The Lexington paper published a letter from an anonymous voter, publicly polling the candidates on the law. Clay, Curd, and Curle promptly endorsed the slave restriction, holding that repeal would flood the country with "refuse and unsound negroes." [3]

Only Robert Wickliffe, Jr., who followed his father's lead in the matter, refused to commit himself. He argued that restrictions upon slave importation did not constitute a suitable subject for campaign discussion. "This issue is not a test question dividing the parties," he charged, "but a question which has hitherto been confined to the halls of legislation." [4] Wickliffe's father, however, was not content with so evasive a response. The Old Duke was sure that the entire discussion was a trick of Cassius Clay's. The fight between Whigs, "in the sight of the enemy," began when the "gentleman, late of Madison County," entered the race, the old man declared. To influence the voters in favor of the newcomer, the antislavery forces had raised the Negro question in the county. The elder Wickliffe said that his enemies were trying to force his only son to "purchase his election at the price of flying at his father's throat." Either young Wickliffe had to denounce his father's stand against the Negro Law or he had to appear opposed to county interests. When the call came, the young man, caught "between the cross fires," could only say, "I will give no pledge about it, but do my duty when called on." [5]

Young Wickliffe's noncommittal position sharply differentiated him from Cassius Clay, who endorsed the law. So well did the issue meet Clay's needs that he had to deny introducing it. "In 1840, I had no share whatever in bringing the subject of slavery before the people," he declared later. But once the question had arisen, he said, his subsequent action followed logically. It was an irrevocable decision. Because he wanted to win a local election and needed an issue to set him apart from his opponent, he took his stand against the expansion of slavery.[6]

Whoever the instigator of the issue may have been, Clay made effective use of the Negro Law in the campaign. With that issue he could appeal not only to the owners of Fayette County's slaves but also to the artisans and small farmers who were in competition with slave labor. From the beginning it had been his devotion to commercial development which made him deplore the presence of slavery in Kentucky. Now he declared that slavery impoverished the state's white population, and before the race was over he had denounced slavery so devastatingly that slaveholders ostracized him.

In his campaign speeches, Clay's purpose was to appear as a friend of the county and its interests. To that end he offered evidence from his previous service in the legislature. "Though a member from another county," he said, "I voted for every measure Fayette ever asked. . . . I was the friend of her literary institutions, her banks, her railroads, her turnpikes." But it was upon the issue of the Negro Law that he spoke most often. He depicted the evils of slavery, basing his arguments upon the injustices of the system, not to the Negro, but to the white. "Give us free labor, and we will manufacture much more than now," he implored. "Slaves would not manufacture if they could; and could not if they would!"

Clay reminded white mechanics that slaves were their rivals: "Negro slavery degrades the mechanic, ruins the manufacturer, lays waste and depopulates the country." He quoted from the Census of 1840 to show that Kentucky's population was not growing at the same rate as that of Ohio, her younger sister. In the previous decade the free state had increased her population by sixty-two per cent, while Kentucky had grown by only thirty-three per cent. "If a free white population *be* itself an element of strength, or the *increase* of population indicates prosperity," he said, "then surely the law of 1833 should stand."

Clay's economic argument against slavery was the weapon he contributed to the arsenal of abolitionism. Other southern opponents of slavery, such as James G. Birney and the Grimké sisters, had directed their appeals to the moral and religious sensibility of slave-owners, only to see erected a moral defense of the institution. But Clay rejected moral preachments as the leading weapon against slavery. "It is not a matter of conscience with me," he said. "I press it not upon the consciences of others."

Instead, Clay tried to prove that slavery was harmful to the skilled artisans who engaged in manufacture. "Every slave imported," he said, "drives out a free and independent Kentuckian." Unless it was checked, slavery would degrade white labor. "The day is come, or coming," he warned, "when every *white* must work for the wages of the *slave—his victuals and clothes—emigrate, or die!*" Appealing to mechanics for their votes, he declared that he favored the white man, "bone of my bone, and flesh of my flesh." The import restriction—indeed, the entire emancipation program—served the interests of the state's embryonic industrial class, Clay claimed, and he vowed his determination to sustain it. "If we pursue the plan proposed by R. W. [Robert Wickliffe], repeal this law, and receive all the surplus vicious slave population which may be thrown upon us, till the whites are thrown into a minority," he admonished, "our strength and influence are gone, our locks are shorn, the star of our glory will have set forever. . . ." [7]

In his efforts to divide the voters, Clay became the object of the undying hatred of the planters. Nearly twenty years later, an obscure North Carolinian, Hinton Rowan Helper, would encounter similar malevolence in defending the yeoman farmer groups against the depredations of Negro competition. But Helper was only enlarging upon the work of Cassius M. Clay, who had long pointed out the deleterious effects of slavery upon another forgotten southern group, the manufacturing artisans. Unlike Helper, Clay was not subject to the charge that he was a lower-class malcontent; because of his background, as well as his courage, he was a more worrisome thorn in the flesh of slaveholding southerners. Like Helper, however, Clay saw nothing good in slavery.

In 1840 he made his position clear: "I declare, then, in the face of all men, that I believe *slavery* to be an *evil*—an evil morally, economically, physically, intellectually, socially, religiously, politically . . . an unmixed evil." [8]

The Wickliffes, in answer, denounced Clay as an abolitionist. The elder Wickliffe, adept at name-calling, referred to Cassius as an "orator of inquisitors, the enemy of Lexington, a secret personal foe, an agitator without spirit, a liar systematically, and an abolitionist at heart." The Old Duke charged that Clay intended the overthrow of slavery in Kentucky. "All that is required to accomplish the emancipation of every slave in the state," the planter said, "is for the abolitionists to get up a war between the slave holders and the non-slave holders." Such, Wickliffe asserted, was Clay's object.[9]

With an explosive issue and expert invective-artists involved, the race was one of the "most exciting canvasses that Fayette had witnessed for many years." While the vote was in progress, Cassius learned that he had a fighting chance to win. After two days of the three-day voting period had passed, he was thirty-two votes ahead of Wickliffe for the third legislative seat. If he held his lead through the final day of voting, he would win. "Help me if you can," he begged his brother Brutus. "I think I can defeat him by hard work." Aided by his brother and other volunteers whom he pressed into service, Clay won the election. He considered it a personal victory. "I, a new-comer, triumphed," he boasted at the conclusion of the count. His success he regarded as another step toward his ambitious goal. "So far," he said, "I had made a good start in my chosen career." But before the end of the session he would revise that opinion. The legislative victory of 1840 was the last election Cassius Clay won.[10]

When the legislature convened and the controversial slave import restriction bill came up, Clay's continued campaign against its repeal made him the target of the proslavery party.

In a lengthy speech he marshalled his arguments in favor of import restrictions. His theme was the evil which slavery worked upon the state's economic development. He read a list of machinery shipped out of New England, and then he rhetorically demanded, "I ask the friends of slave labor how long shall we wait till we shall be able to supply Europe and the world with such things of manufacture?" How long, he asked, "before Holland will send to Kentucky for grist-mills?" Kentuckians had waited two centuries for that, he said. "Shall I, then, be taunted with Yankee feeling because I would dispel the lethargy which rests upon our loved State?" As the advocate of a Kentucky System, he depicted the opportunities for profits in belching smokestacks and droning machines. Four decades before Henry W. Grady would preach the gospel of a southern factory economy, Clay had seen the vision of a New South.

Continuing his arraignment of slavery, he declared that the servile labor system despoiled the fertility of southern soils: "Ignorance and carelessness, which are necessarily combined in the slave, make his the most slovenly and wasteful of all labor." Moreover, the presence of the slave not only degraded labor; it also reduced the managerial skills of the owners. No one in Kentucky should be surprised, he said, that the North was "radiant with railroads, the channels of her untold commerce, whilst the South hobbles on at an immeasurable distance behind." [11]

As spokesman for the Kentucky System, his own scheme for attaining his ambitions, Clay sought to represent the state's non-slaveholding masses against the relatively few slave-owners. When the proslavery spokesmen stated that "white laborers are slaves," Clay declared that he would oppose them "in the name of five hundred thousand free-men of Kentucky." Here was his potential following, and he worked to become its political leader. But he faced difficulties, chief among them being the accident of his time. He had arrived too late upon the state's political stage. Earlier he might have aroused an interest in industrial development; he might even have, without fear of retaliation, described slavery as an evil. He knew that his kinsman and model, Henry Clay, had fathered the American System, as well as numerous compromises with antislavery politicians. By 1840, however, fear of abolitionists had driven southern leaders to demand conformity upon the slave question. There should be no discussion of it: "We shall play into the hands of northern fanatics by this course," they seemed to agree. For Cassius Clay, who sought a career by following in his cousin Henry's footsteps, it was a disastrous development.[12]

The assaults of his political opponents were not long in appearing. Clay's attack upon slavery gave the younger Wickliffe the issue he needed for the campaign of 1841. In mid-April, some months before the race would normally begin, the Young Duke spoke against Clay, associating him with the abolitionists. Wickliffe "attacked my course in the legislature violently," Cassius reported to Brutus. "There is much excitement arising, and my friends demand my entire devotion to the canvass." But the long-suffering Brutus, mainstay of the family, complained that Cassius ought to attend to his personal business and meet his pressing financial obligations. "I am only desirous to run this time for the legislature," Cassius responded. After 1841, he promised, he would have no temptation to desert his business. But he regarded the election as all-important: "It is the crisis of my life and I must meet it or fall." [13]

Cassius Clay met his crisis with characteristic ferocity. Less than a week after Wickliffe had commenced war upon his legislative record, the two men had made a challenge for a duel. "The immediate cause of the quarrel arose out of Clay's difficulty with old Mr. Wickliffe," explained John C. Breckinridge, prominent Kentucky politician. On the night of April 24, a heated and bitter debate occurred in Lexington between Clay and Wickliffe, Jr. Clay's father-in-law, Warfield, reported that it was "likely to result in a serious manner." Wickliffe had reiterated his charges that Clay, in admiring Yankee prosperity, was therefore an agent of the Yankee abolitionist movement. In denouncing his opponent, Young Bob made such insulting personal allegations that Clay promptly challenged him to settle the matter by physical combat. "It is generally thought that Mr. Clay was too hasty in making young Mr. Wickliffe responsible for his father's conduct," reported a Lexington lady, "though Mr. Clay says he is acting altogether on the defensive." [14]

Protesting his innocence, Cassius left Lexington with his seconds, a fellow officer in the Kentucky militia, Major William R. McKee of Garrard County, and veteran Whig politician Thomas A. Marshall of Frankfort. Together they journeyed to the Indiana duelling ground where they would meet the Wickliffe party. Clay's wife, staying with her parents, was ignorant of the impending fight. "Mary Jane knows nothing of our fears," her father assured Brutus, "nor will she if we can prevent her from such disastrous intelligence." One of Clay's brothers-in-law remarked that Cassius was determined to receive satisfaction. "Wickliffe must back out or there must be a fight." [15]

Cassius had sharpened his shooting eye for such an eventuality. So skilled had he become, it was reported, that he could sever a string with a bullet from his pistol three times out of five shots at ten paces. Yet on May 13, 1841, in the duel with Wickliffe, the contestants exchanged three shots at ten paces without effect. Cassius demanded another round, but the seconds intervened and ended the encounter. Upon hearing the news, a Lexingtonian commented, "Robert Wickliffe and C. M. Clay's duel was settled without blood. Tho I think there is *bad blood* left." No apology was made by either man, and Clay reported, "We left the ground enemies, as we came." [16]

Some days later a friend teased Cassius about his marksmanship. "Why is it," he wanted to know, "that you could cut a string at ten paces three times out of five, and yet miss Wickliffe's big body three successive shots at the same distance?"

"Oh," drawled Clay, "the damned string had no pistol in its hand." [17]

Cassius missed his mark, and he also failed to win re-election in the 1841 campaign. As he had predicted, the race was the crisis of his life. It was his final effort to advance politically in the Kentucky Whig organization. He did not put aside his ambitions, but sought the support of artisans and laborers rather than that of slaveholders. He had won the enmity of the minority dominant in the state, and he protested that they fought him with illegal weapons. He claimed that he had won the election, only to lose it through fraud. "I most solemnly reiterate," he said, "that I believe that I received a majority of the legal votes of Fayette County." But, he declared, he had been swindled by the slave party—"every judge of the election in all the precincts being against us." The "damning infamy" which had "all at once ruined such seeming prosperous career," he said, was that he had "turned TRAITOR TO SLAVERY!"

But he remained loyal to the white non-slaveholders of Kentucky and called upon them to vindicate him at the polls. He had sacrificed a promising career, he said, for the "*six hundred thousand free white laborers of Kentucky!*" They were the people "against whose every vital interest slavery wages an eternal and implacable war!" It was for their welfare that he had repudiated the planter class and risked his career. "Yes, these are the men, the great majority of the people of Kentucky," he declared, "whose interests, in 1841, I swore I would never betray—for whom I then fell. . . ." Clay proposed a political solution; he appealed to the free whites, who had no vested interest in slavery but were its victims, and who had the power to reinstate him. "How long, my countrymen," he implored, "seeing you have the power of the ballot-box, shall these things be? . . . Will you not at last awake, arise, and be men? Then shall I be delivered from this outlawry, this impending ruin, this insufferable exile, this living death!" [18]

Cassius Clay staked his career upon an audacious gamble that he would succeed in separating the six hundred thousand from the few thousand who owned slaves. If the artisans and manufacturing laborers would vote in their own interest, he told them, then they would defeat the masters and place him once again in the profession for which he had prepared himself. That time had not yet come, however, and the political exile which confronted him was long and dreary. Before the laborers would vote for their economic interests, he would have to educate them to regard slavery as inimical to them. He would also have to face the difficult race problem. Many of the non-slaveholding whites may have agreed with him that the institution worked to their disadvantage, but the alternatives appeared distasteful. A strong race sentiment tended to keep the white population united in support of any system which would keep the Negroes under control. With his task clear, the next phase of Clay's activity was the thankless, unrewarding position of theorist and propagandist for the Kentucky System.

CHAPTER IV

PUBLICIST AND BOWIE KNIFE EXPERT

CLAY slowly matured into a capable defender of his cause. He was in no hurry to begin his comeback, and for a full year he kept out of local politics. As he had promised Brutus, his first concern was for his business. Complaining that he was much trammeled by securityship, he forced himself into the confining routine of his mill enterprises and his farm. But he did not lose sight of his ambitious objective to win a career in public life, and during the year of retirement he coldly calculated his plan of attack. He had recognized that success depended upon his educating a majority of Kentucky voters to accept his contention that slavery harmed them. Because his ideas were distasteful to influential Kentucky slave-owners, he faced the problem of finding ways to reach his audience. Free white workers in the South, he charged, were "*barred by despotic intolerance from receiving any light by which they can know their rights, and free themselves from the competition of slave labor, which brings ignorance and beggary to their doors.*" To bring that light and to organize political opposition to slavery, Cassius Clay began a publicity campaign. Working in a slave area where converts counted most, he had an unusual opportunity to affect the outcome of the abolition crusade.[1]

Clay emerged from his retirement to take part in debate

over repeal of the Negro Law. Early in 1843 the lower house of the state legislature repealed the slave-import restriction, and while the Senate deliberated the measure, Clay broke his long silence. Through the columns of the *Lexington Intelligencer* he reiterated his argument that there was a direct relationship between the political power of the slave party and restraints upon manufacturing in Kentucky. To insure the continued domination of the proslavery politicians, "Everything of value would be given up," Clay said; "our free white laborers are to be driven out; our manufactories, already too inconsiderable, are to be destroyed; our cities are to crumble down; our rich fields to grow sterile." Yet, in defiance of the public interest, the legislators had repealed the law and threatened the state with an "influx of foreign degraded slaves." The Negroes who would enter the interstate slave market were the unwanted castoffs, he declared: "House-breakers, poisoners, rogues, perpetrators of rapes and midnight murders." [2]

With vituperative disgust Clay described the Negroes who would be brought into Kentucky. Behind his vehemence there was a reason. White men critical of Clay might charge that he advocated emancipation out of love for the Negro, and he intended to meet the argument in advance. He always maintained there was no doubt that Negroes were inferior to whites, though he did at one time suggest that the Negro race could be improved. And it was not difficult for Clay to avow a dislike for the black man; even in his private communication he habitually mentioned Negroes with contempt. "I have studied the Negro character," he wrote a few years later. "They lack self reliance—we can make nothing out of them. God has made them for the sun and the banana!" Clay insisted that he fought the slave system, not because he loved the Negro, but because he wanted to assist the white. "If we are for emancipation," he explained, "it is that Kentucky may be virtuous and prosperous. If we seek liberty for the blacks, it is . . . that the white laborers of the state may be men and build us all up by their power and energy." He said he favored an emancipation program "not because the slave is *black* or *white*—not because we love the black man best, for we do not love him as well, . . . but because it is *just*." But however much he protested, he was never able to escape the derogatory taunt that he acted out of love for the Negro. His enemies circulated the familiar toast, already trite in the 1840's, which reportedly originated at a " 'darky' celebration down South": "Massa Kashus M. Klay—de friend ob de kullud poppylashum:—aldough he hab a wite skin he hab also a berry brack heart: which titles him to the universal 'steam ob dis 'sembly." By such gibes, the slavery party reduced the effectiveness of Clay's efforts without ever answering his arguments.[3]

Clay, however, kept his campaign on a level of factual argumentation. Later in 1843, he published in Horace Greeley's *New York Tribune* an antislavery tract which Greeley extracted and circulated as a pamphlet entitled "Slavery: The Evil—The Remedy." For its prologue, Clay repeated his reasons for opposing slavery, referring to census figures to show that slave states were far behind free states in education and in mechanical progress. But he devoted the majority of the letter to an examination of emancipation as a safe remedy for the evil. Writing in a northern journal with a message designed for southern readers, he repudiated the "higher law" moral absolutism which motivated the abolitionists. Some of the Yankee reformers, represented by the forthright William Lloyd Garrison, demanded immediate emancipation, regardless of law or tradition. Theologians of humanitarianism, they resorted to sanctimonious appeal to a law higher than man-made statute. Such a program, threatening southern customs, presented an obstacle which Clay had to overcome if he were to win the ballots of southerners.

He met that hurdle, as he met the others, with characteristic directness. He was a politician, not a moralist; more important, he was in Kentucky, not in the North. A program which might be politically feasible in a free state would not necessarily be acceptable in a slave area. Indeed, any antislavery program would be suspect among slave-owners. "You cannot properly appreciate my position," Clay told the Ohio abolitionist Salmon P. Chase. "All abolitionists are not like yourself, moderate, reasonable men. They are, many of them, incendiaries; with such neither I nor my people can have any consideration." Therefore he categorically denied the "higher law" doctrine. "I cannot agree," he said in his message to the *Tribune*, "that there is any law superior to that of the federal Constitution." His would be a moderate course, entirely legal, with an appeal to the ballot its predominant feature. A constitutional campaign against slavery, Clay reasoned, would be the only kind of attack acceptable to southerners. So long as the constitution of Kentucky sanctioned slave property, then the law must rule; but as the constitution set forth the means of its amendment, he claimed the right of advocating its change. In a republic, the majority had the power to change any law. When he amassed a majority in Kentucky, he would liberate the slaves by legal means. Because Clay respected the law and tried to change it, his plan differed basically from that presented by "higher law" abolitionists, who disregarded man-made law and demanded emancipation as a humanitarian service to the enslaved. A constitutional appeal directed to the economic self-interest in non-slaveholding whites would be more effective in Kentucky than the emotional immediatist demand. Primarily a political theorist, Clay had worked out a plan which might restore his lost career.

He based his appeal upon statistical evidence against slavery, and he tried to prove deleterious effects of slavery upon the state's welfare. "I should be glad to have occasionally documents from you should such be printed, showing the *comparative wealth, arts, sciences, numbers*, etc., of the slave and non-slave states," he wrote his friend Chase. "These are better arguments than invective. The one awakens at least a hearing. The other shuts the ears as well as the conscience." In time he would amass a significant body of evidence to show how slave states lagged behind free states, and he would infer that slavery was the cause of it.[4]

From the beginning, Clay's publicity program made an impression upon observers in Kentucky and abroad. A Lexington lawyer estimated that there were "thousands of Abo." in Kentucky, and predicted that Clay would win a following among them. Across the river in Cincinnati, the editors of the *Gazette* praised him: "Cassius M. Clay, of Kentucky, has denounced slavery in stronger language than any man we know." Farther away, other people acclaimed Clay. Lewis Tappan, secretary of the American and Foreign Anti-Slavery Society, assured Clay that his tract was "read with peculiar interest at the North"; and an amateur poet hailed "Cassius M. Clay, Emancipator," as

> Star of Kentucky, and voice of the Free!
> Now we shall hear again
> Liberty's awful strain
> Rolled to the Southern plain
> From the North Sea.[5]

The growing admiration for Clay's program frightened the slaveocracy, who worked to limit his influence. They used the weapons of fear and intimidation against him and his potential following. Although he urged Kentuckians to "face opposition—and be the better for it," not many possessed his courage. Clay himself did not bow before the threat of violence. "I knew full well that the least show of the 'white feather' was not only political but physical death," he said, and he kept his senses alert to detect signs of attack. He did not hesitate to speak his mind in public, and everywhere he went he carried his pistols and bowie knife. Soon after the publication of his *Tribune* letter, he had an opportunity to use them.

Since 1841, when he had abandoned local politics as a candidate, Clay had continued to support Whig principles and office-seekers. In the summer of 1843 he took an active part in the election for representative from Kentucky's Eighth Congressional District, which included the Bluegrass area. Cassius and his friends—"laboring men mostly," he said—supported the polished Garret Davis of Bourbon County, an ardent Whig and a close friend of Henry Clay. Cassius' action reopened the unhealed wound between himself and the Wickliffes, for Robert, Jr., was the Democratic nominee. Both sides indulged in bitter personal invective, and the matter came to a climax just a few days prior to the election.

Young Wickliffe charged that the Whigs had been guilty of dishonesty in the recent revision of the congressional districts. A group of Whigs, Wickliffe declared, had met in Frankfort to arrange the districts so that Davis would be brought into the Fayette district. Such a procedure might enhance Whig chances in the Lexington area. Davis promptly denied the allegation and called upon his opponent for the proof. Wickliffe, in turn, cited as his source a man who lived in Woodford County, but Whigs elicited from the man a statement that the story was "a lie—yes, a damned lie." Cassius was aware of this denial, and he had the courage to proclaim his information.

On the morning of August 1, Wickliffe spoke in Fayette County, at the village of Delphton. Davis was not present, but Cassius was in the audience, armed with a well-honed bowie knife and prepared to represent the Davis candidacy. Wickliffe, in the course of his address, repeated the charge that the Whigs had gerrymandered Davis into the Fayette district. When he did so, Clay arose and broke in upon his discourse in what he later termed a "calm and respectful manner." In his stentorian voice, he interposed: "Mr. Wickliffe, justice to Mr. Davis compels me to say . . . that it was a damned lie." But, he continued, "I have no intention to interrupt you; go on." The speaker proceeded, and there was no trouble.

At three o'clock the same day, Wickliffe had another speaking engagement in the county. He appeared at Russell's Cave Springs, a popular meeting place and picnic ground a few miles north of Lexington, where a creek flowed from a cavity in the earth. Clay followed Wickliffe to defend the Whig case. Another member of the audience was Samuel M. Brown, a New Orleans post-office agent, and a noted fighter endowed with a quick temper. An ex-Kentuckian, he had returned to Lexington on vacation, and went to hear the campaign speeches. Ever afterwards, he regretted that he had attended.[6]

In his afternoon address Wickliffe repeated his indictment of Whig leaders, ignoring Clay's remonstrance of the morning. When Young Bob came to that part of his speech, Clay again arose and interrupted the speaker. "Mr. Wickliffe," he boomed in his powerful voice, "I have listened to you with great patience, and shall hear you through; I do not wish to interrupt you; but justice requires that . . . I should state the opposite side of the question." Clay then presented his evidence, branding the gerrymander accusation a Democratic fabrication. At that point, Brown, who stood near Clay, entered the argument. "Sir," Brown shouted heatedly, "it is not true." Clay turned to the new disputant, and replied, "You lie." Brown yelled, "You are a damned liar," and rushed at Cassius.

Clay held a leaded horsewhip in his hands, and he rained blows upon the head of his assailant. As he struck Brown, bystanders seized Clay from behind and separated the combatants. Quickly, Cassius threw away the ineffective whip and drew his bowie knife. At the same time someone in the crowd handed Brown a pistol. Clay turned to see Brown facing him, brandishing a pistol in his hand and shouting, "Clear the way and let me kill the damned rascal!"

Without hesitating, Cassius charged at his opponent. Brown held his fire until Clay was almost upon him. At point-blank range he fired, and then Cassius closed in, hacking and stabbing Brown's unprotected face and head. He "cut away in good earnest" until men in the crowd could separate them again. Brown had lost his right eye; his left ear and a piece of his skull were "lopped off"; and blood streamed copiously down his face from a nose "cleft in twain." Senseless, Brown sank to the earth.[7]

When the crowd prevented Cassius from using his knife to do further damage, he picked up his adversary and carried him to the edge of the field. He tossed Brown's unconscious body over a bluff, and it rolled down into the creek. Leaving Brown for bystanders to fish out, the victor, breathless and exhausted, went to a nearby farmhouse to inspect his own wound. There he discovered that Brown's ball had penetrated his clothing but had lodged in the silver knife-case which he wore under his coat. Miraculously he had emerged from the fray with only a red spot over his heart.

After the fight Clay's reputation as a bowie knife expert spread abroad. Many of his political opponents would take a lesson from Brown's unfortunate experience and steer clear of the flashing blade. Clay served notice on the proslavery party that he would defend his right to differ with them, and they came to respect his fighting abilities. The way he handled his knife, a witness marvelled, "was tremendious." [*sic*] [8]

To question the responsibility for Clay's efficient bladework, the Fayette County Circuit Court indicted him for mayhem. As his defense attorney he engaged his honored kinsman, Henry Clay. The courtroom overflowed with admirers of the Gallant Harry's deft legal tactics. He did not disappoint his following. His case, one reporter commented, was "eloquent and scathing." He made no effort to defend Cassius' political views to the jury but sought to prove that his client had acted upon the defensive and was therefore innocent. "You are bound, on your oaths, to say, was Clay acting in his constitutional and legal right?" he demanded of the jury. "Was he aggressive . . . standing, as he did, without aiders or abettors, and without popular sympathy; with the fatal pistol of conspired murderers pointed at his heart, would you have him meanly and cowardly fly? Or would you have had him to do just what he did . . . ?" Then, turning toward Cassius, with his voice "broken but emphatic," and raising himself to his full height, the master legalist said, "And, if he had not, he would not have been *worthy of the name which he bears!*" Adroitly playing upon the jury, Henry Clay, keeping clear his record of not losing a criminal case in the last forty years of his practice, won acquittal for his client.[9]

His cousin's masterly legal brief, as well as his own competent bowie knife surgery, provided Cassius Clay with valuable publicity. Admirers all over the country expressed sympathy for his position among so dangerous a populace. A Cincinnati editor eulogized Clay as the defender of the rights of the working masses. "The broad spirit of philanthropy which is in the man, his fearlessness whenever human rights need a defender," would, he remarked, melt all opposition. Resistance to Clay's program was "superficial, skin-deep," and did not come from the "thinking man; the mechanic, farmer, and day laborer.—They are with Mr. Clay." The editor praised him as the most effective of the antislavery propagandists; Clay was doing more, he said, "in his sphere, than any of us in the free States, to rid the country" of slavery.[10]

In the first year of his publicity campaign, Clay had employed what means were available to address Kentucky voters. He utilized the columns of the local Whig paper until the editor would no longer accept his contributions. Then he turned to the metropolitan press, expecting that the local journals would reprint his articles, but that proved to be a false hope. "My letter to the Tribune has not been republished in the slave states and I have not been able to force it into the press by any means," he reported to Chase. "The press is monopolized by the slave-holders and the people receive no light and are filled with prejudices carefully instilled into them." He wanted a press of his own but admitted that it was not immediately possible. "I fear that . . . until you throw more light among the laboring people in the slave states, . . . a free press could not stand against *violence* here. The slavery men are united and can move in mass. . . . But I think the time not far distant when this can be done, and then the cause will go on." But before that time came, the election of 1844 provided Clay the opportunity to express his views to a wider audience.[11]

CHAPTER V

AN EMISSARY FROM COUSIN HENRY

IN 1844, after years of political effort, Kentucky's Henry Clay finally received the Whig nomination for the Presidency. Famed as the creator of the American System and for his long career as a statesman, the keen-eyed, sharp-tongued Clay was a widely admired American. His chief opponent, the Democratic choice, was the comparatively unknown Tennessean, James K. Polk. Although for years Polk had been active in state politics and had served in Congress, Whigs alleged that he was a newcomer to national affairs. Derisively they chanted, "Who *is* James K. Polk?" Led by Polk, the party of Andrew Jackson vociferously demanded westward expansion and the annexation of the Republic of Texas. On the ground that annexation was unconstitutional, antislavery Whigs readily took up the challenge.

To complicate the party division over Texas, however, there was a third party in the field. The new-born Liberty Party, headed by ex-Kentuckian James G. Birney, was a reform party which emphasized immediate emancipation of slaves. Although the Liberty Party had polled only seven thousand votes in 1840, northern Whigs—in particular New Yorkers Millard Fillmore, Washington Hunt, and Thurlow Weed—feared that it might encroach still further upon their party vote. To weaken the Birney following in the North, they worked to make the Whig Party acceptable to abolitionists. They needed a person who subscribed to the economic principles of the party and who would at the same time have an influence upon the antislavery voters. To fill those requirements, northern Whigs invited Cassius M. Clay to tour their section on behalf of their candidate.

To all appearances, Cassius was the ideal choice. He was a proponent of the American System, and in addition he had a wide antislavery reputation. His emancipation articles, written for Kentuckians but published in the northern press, had attracted national acclaim. "Your writings and corresponding deeds on behalf of the down-trodden have endeared you to the humane and liberty-loving everywhere," abolitionist Lewis Tappan of New York City assured Clay. Moreover, in January, 1844, Cassius had emancipated his own slaves, an action which made him more acceptable to Yankee abolitionists. Furthermore, Clay's courage in facing threats and drawn pistols, as well as his southern background, made him an attraction to curiosity-seekers. That he was a relative and a close associate of the candidate, Henry Clay, was another asset.[1]

Cassius had also won the confidence of other Liberty Party organizers, such men as Henry B. Stanton of Boston and Salmon P. Chase of Cincinnati, by his strong stand against Texas annexation. On May 13, 1844, at a Lexington public debate on the Texas treaty, he had answered Thomas F. Marshall, his old Whig colleague now turned Democrat. Tall and slim, with twinkling eyes and a heavy black beard, and about the same age as Clay, Marshall had spoken as an advocate for the Democratic platform, which demanded the "re-annexation" of Texas. After hearing Marshall for three hours, Cassius Clay arose in rebuttal. Massive and cleanshaven, his appearance as well as his argument differed from Marshall's. Unlike the partisan Democrat, Cassius posed as a public-spirited neutral. He spoke not as a Whig, nor as a representative of Henry Clay, but as a "citizen of Kentucky, and of the United States, a southerner by birth, association, and feeling." From that position he opposed the Texas treaty as "revolutionary, mad, and fatal to my country." If the Senate could unite foreigners to the confederacy today, he said, they could do the same tomorrow. They could, he added with a wry grin, "merge our very nationality into the first despotism which shall be able to insinuate gold enough into their pockets to outweigh the patriotism in their bosoms." If the Constitution were to be thus lightly set aside, he warned, then Americans had lost their freedom. "If we are at the whim of a president and fifty-two senators," he asserted, "then we are slaves and not free." That was the important issue to Cassius Clay, and he repeated it again and again. "It is not, with Texas and a slave-holding Senate, whether we assent to slavery," he shouted, "but whether we ourselves shall be slaves!" [2]

His efforts were not lost on northern Whigs, who welcomed his assistance. "We have . . . to fear . . . the Abolition vote," Millard Fillmore told Thurlow Weed. "Cassius M. Clay can do much to aid us." From Niagara, Whig Congressman Washington Hunt praised Clay's effectiveness as a party spokesman. "He has a way of presenting the Texas question in clear and striking points of light," he said, "and he can do much good in some of the Abolition counties, such as Madison." Fillmore jubilantly reported that the Kentuckian had consented to make a campaign journey in the North. Clay would speak in Rochester and in Boston, he said, and then would devote "the rest of his time till election in attending meetings as we shall think best . . . no time is to be lost." [3]

Cassius' decision to participate in the campaign caused an immediate reaction among the abolitionists, indicating that his task would not be a simple one. As soon as he announced his intention to support Henry Clay, Liberty Party men declared war upon him. How, they wanted to know, would

Cassius justify endorsing a slave-owner? He answered that he would support Henry Clay despite his slave property. "Mr. Clay is indeed a slaveholder—I wish he were not," Cassius admitted. "Yet it does not become *me,* who have so lately ceased to be a slaveholder myself, to condemn him." Just this once, and for the last time, he promised, he would vote for a slave-owner. Soon public opinion would reject such a candidate. After 1844, he predicted, "no man . . . should be deemed fit to rule over a Republican, Christian People," who violated, by holding slaves, the only principles "upon which either Christianity or Republicanism" met the "test of philosophical scrutiny." [4]

Despite Cassius' excuses, his support of Henry Clay brought gibes from Liberty Party men. Gerrit Smith, of Peterboro, New York, a spokesman for the Birney party, ridiculed Clay's explanation. "We have a class of Abolitionists who are called the 'just-this-once-men'," Smith said. "Next Autumn," he told Clay, "will witness your last sin against your enslaved brethren." But Smith's sarcasm did not affect Cassius. The clash merely emphasized the rift between him and the humanitarian reformers. He made no effort to hide his opinions; he repeatedly rejected the moral harangues of religious abolitionists. "I have not at any time assumed to be better than other men," he said, "and whilst I profess to be open to the sympathies of our nature, I have never set myself up as a philanthropist." Clay did not regard emancipation as a means of wiping out a sin. He expected it to be attended by significant economic and social gains and by the general improvement of the public welfare. He also anticipated a political career from his interest in it. He was not, unlike Birney, an abolitionist who used the ballot box to achieve humanitarian reforms and to whom emancipation was the chief end sought.[5]

Clay and Birney did agree, however, on the urgency of the Texas issue, and Cassius proposed that the Whig Party take an uncompromising stand against slavery and fight the election upon the division "*Polk, Slavery, and Texas,* and *Clay, Union, and Liberty.*" By doing so, the party might lose several slave states which were considered certain, but he anticipated winning enough northern states to compensate for them. The time had come, he declared, to stand or fall on the basic issue. "It is in vain to put off the evil day," he said in a communication to the *New York Tribune.* "Slavery or liberty is to be determined in some sort this coming election." Cassius wanted to make a clear division between the parties over slavery, not over Texas alone.

The assertion that slavery was incompatible with the Whig program was consistent with Cassius Clay's political beliefs. Slavery, he had proclaimed, was merely the watchword for a southern economic system that was inimical to industrial prosperity. In a widely published letter to a New York Whig audience, he tried to steal the Liberty Party's program by connecting slavery with debt failure, which was the nightmare of the bankers. "Save us from disgrace and ruin," he prayed. "Elevate us among nations to that post of honor which we once held and from which slavery and repudiation—twin brothers—have dragged us down." [6] It was not slavery in itself that had dragged the nation down, but the combination of slavery and an economic bugaboo. In the aftermath of the panic of 1837 some slave states had repudiated their debts, and Cassius connected an economic practice with a social institution. He appealed to northern economic interests, whether of Whig or of Liberty Party leanings. Long before the national Republican Party was organized, he suggested that the controversy over slavery might serve to rally divergent northern groups in support of the New York–New England industrialists and financiers.

Before he departed on his speaking tour, Cassius had two duties to perform. The first was his obligation to the state as a militia officer. As a colonel, Clay had been invited to command the state's summer encampment, but some of the officers refused to serve under his command. When his friends arose in his defense, the opposition was squelched, and in the first week of July, Cassius assumed command of Camp Hart in Woodford County, near Versailles.[7]

The encampment was typical of militia affairs of the time; it was more holiday than military service. On July 4, there was an oration—to show his magnanimity, Colonel Clay invited his recent debate opponent, Thomas F. Marshall—and the firing of a twenty-six-gun salute. Following the patriotic ritual, the governor of the state, clad in resplendent uniform, inspected the parade. For the remainder of the encampment the citizen-soldiers competed in target-firing, played games, and dined upon roast venison. At the conclusion of the week, Colonel Clay received the "warmest gratitude" of his command, and, in the sycophantic language of junior officers, his subordinates praised the "firm and zealous manner in which he has caused his orders to be obeyed." [8]

When the week of camp was over, Cassius called upon the Whig candidate, Henry Clay, to receive blessings and instructions before he began his tour. Henry had disagreed with Cassius' interpretation of his own teachings and differed with him on the Texas annexation. Henry Clay had straddled the issue so satisfactorily, in fact, that he satisfied many southerners and would even take Tennessee from Polk. But despite his differences with the extremist Cassius, Henry Clay was willing to use him. He had an opportunity to become President, and he would take whatever steps were necessary to win. When Cassius showed him the invitations from the North and explained the mission that New York Whigs had planned for him, Henry Clay gave his consent. It would not be long, however, before Henry would disown his outspoken cousin.[9]

His plans to travel through the North forced Cassius to

put his devotion to politics to a severe test. If he made the trip, his wife would accompany him, and they would have to leave their children, now two daughters and a son. Both parents feared to trust their family to the servants. Only the year before, Cassius Marcellus, Jr., then four years old, had died, and Cassius suspected that he had been poisoned. Among his slaves (those whom he could not emancipate because his father had entrusted them to him) was a nurse named Emily. Cassius announced that he had "every reason to believe" that Emily had poisoned his son with a "deadly poison called arsenick." [10] To leave his children was, therefore, a difficult choice. Cassius may have been unrelentingly hostile to his enemies, but he was a sympathetic parent, solicitous of his children's welfare. When they were ill he would not leave home, not even to go to the far side of the White Hall estate from the big house. He would stay at the bedside of the sick child, to entertain him and to care for his needs. Clay's allegiance to the Whig cause triumphed, however, when he decided to campaign as an emissary from Henry Clay. Leaving the children with relatives, Cassius donned his campaign togs—a brass-buttoned blue suit, and he and Mary Jane departed for the political hunting grounds across the Ohio.[11]

In 1844, political campaigning presented unusual physical hardships, and the first requirements for a speaker were a powerful voice and a resilient physique. For that reason, Cassius Clay was an admirable campaigner. It was the first extended speaking tour he had made, but his journeys to the East had prepared him for what he was to encounter. To Mary Jane, however, the trip was a revelation.

For a part of their journey into Ohio the Clays went by rail, a method of travel which presented its own torments. Across country, they resorted to a horse-drawn buggy and bumped over the Ohio pikes, escorted from town to town by enthusiastic Whigs. In late August, they reached Jefferson,

Ohio, where Cassius had a speaking engagement. There the procession received a royal welcome of "several Buggies and a Wagon with flags and a band of Music." Joshua Giddings, an Ohio abolitionist congressman whose home was in Jefferson, met the Clays and accompanied them into the town. Mrs. Clay, worn out by the rough travel, asked to be taken to a hotel room. "I asked for a private room, they carried me to one," she reported, "but I might as well have been carried into the Public Dining Room. . . ." The wife of a celebrity, Mary Jane was learning, had to forego the luxuries of rest and privacy.

After Cassius had spoken in Jefferson, the party headed for Paynesville, where they arrived at four o'clock after a hard, all-day jaunt. There Cassius again met the crowds of inquisitive Ohioans. Wherever they went, people pushed to see and to touch him. "You see them in flocks peeping in and whispering," Mary Jane marvelled. She heard one hostess admonish her son, "Now Johnny, don't get to fighting, remember we've got President Clay in the house." With more than one Clay involved in the campaign, the confusion multiplied.[12]

The fact that both candidate and campaigner bore the same family name caused Henry Clay many worries. The very features about Cassius which made him an acceptable emissary to northern abolitionists harmed Henry's chances in the South. As Cassius had undertaken the mission of converting Birneyites to Clay, he depicted his kinsman as an emancipationist. "I believe his feelings are with the cause," Cassius said in a letter published in the *New York Tribune*, and he added that "the great mass of Whigs are, or ought to be, anti-slavery." For that statement, the candidate repudiated his emissary. "Mr. C. M. Clay's letter," Henry explained, "was written without my knowledge, without any consultation with me, and without any authority from me. . . . He has entirely misconceived my feelings."

Henry Clay reported in confidence to Joshua Giddings that he regretted the necessity of disowning his kinsman, but he feared the loss of four slave states if he did not. His advisers had warned him that he might not even carry Kentucky. Meanwhile, Henry cautioned Cassius to restrain his antislavery ebullience. "As we have the same sirname, and are, moreover, related, great use is made at the South against me, of whatever falls from you.—There, you are even represented as being my son: hence the necessity of the greatest circumspection."

Henry's efforts to quiet Cassius illustrated the Whig dilemma. "At the North," Henry said, "I am represented as an ultra-supporter of the institution of slavery, whilst at the South I am described as an Abolitionist; when I am neither the one nor the other." As Cassius had pointed out, the sectional split could no longer be ignored; he wanted to solve the problem by expelling the proslavery members. But Henry Clay, the master compromiser, expected to win the victory by avoiding a showdown statement. Cassius, impetuous and outspoken, would never comprehend his cousin's political agility. The compromising candidate now began to express doubts that the uncompromising Cassius would succeed in his mission. "After all," he counselled Cassius, "I am afraid you are too sanguine in supposing that any considerable number of the Liberty men can be induced to support me." [13] But Cassius was young, exuberant, and optimistic. He determined to continue his campaign, and in September, with Mary Jane at his side, he energetically spoke his way through Ohio and Michigan.[14]

For a month the Clays campaigned in the Buckeye State with Governor Tom Corwin and staged rousing rallies reminiscent of 1840. Corwin, a master of the comic political monologue, delighted the audience with his witticism. "And *who* have they nominated?" he would demand, in his drawling voice. "James K. Polk, of Tennessee?" Then, wagging his head slowly from side to side in mock amazement, he would ask,

"*After that*, who is safe?" But while the governor sought to create the illusion that Polk was an unknown, Clay chafed at Corwin's religious fervor. "What struck me as most remarkable . . . was his indulgence in 'whining, canting, and praying in his speeches,' " Cassius recalled a few years later. "I have been in the furor of revivals, and the wild enthusiasm of the bivouaced camp-meeting, and never did unctious Methodist parson move me to tears like the 'inimitable Tom!' " Corwin quoted Scripture to the Ohio audiences until Cassius squirmed in horror at the blasphemy. But when he complained about it, the governor responded that "no people were so conscientious and devout as these . . . Abolitionists. . . ." Before the campaign was over, Cassius would learn that Tom Corwin was right.[15]

While Corwin combined humor with liberal dashes of antislavery homiletics, Cassius Clay offered straight campaign fare with a humor all his own. His speech at Cleveland was typical of others in Ohio, where his unvarying theme was the Texas annexation. Henry Clay, he said, would not accept the Lone Star Republic unless it came in "without dishonor, without war, with the common consent of the Union, and upon just and fair terms." As those conditions were unlikely, Cassius asserted that in effect Henry opposed the expansion. He pointed out that if the Whigs won, there would be no immediate annexation, while if Polk won, Texas would enter within twelve months, "with dishonor and war and ruin."

To such an interpretation of his views, Henry Clay offered no protest. But when Cassius appealed for the Birney vote, Henry feared that his cousin would drive off the southerners. Ignoring Henry's repudiation, Cassius sought abolition support by reiterating his claim that Henry approved his emancipation activities. "I am a practical abolitionist," he told the Clevelanders. "The destruction of the whole system of slavery is what I seek above everything else . . . with this object before me, I earnestly advocate the election of Mr. Clay as an instrument for the accomplishment of that great purpose." Cassius also alleged that Henry agreed with his opinions "in substance." Coming from one who was regarded as an emissary from Henry, it was an effective recommendation to those who had not read the candidate's disavowal of the campaigner. Moreover, as a southerner, Cassius claimed to know the secret schemes of slaveholders. They hoped to succeed, he told the Ohioans, by the indirect aid they would get from northern abolitionists who diverted Whig votes. Thus, he declared, antislavery Yankees who voted for Birney would aid slaveowners, the group they anathematized. It was a "most unholy alliance," Cassius concluded.[16]

With his exposition of the Texas issue, and with his personal appeal for the antislavery vote, Cassius Clay and his wife traveled throughout the Midwest and then entered New England. To Mary Jane the trip had become an exhausting series of similar scenes. Always there were escorts to meet them on the highways, torchlight parades to disturb the night, banquets and interminable speeches to endure, each an echo of another. Cassius, handsome and earnest in his dark-blue suit with its polished buttons gleaming in the lamplight, drew applause from partisan gatherings when he endorsed Henry Clay. He was an effective stump speaker and enjoyed the give-and-take of a political rally. Again and again he drew laughter as he matched sallies with his hearers. Far from the raised eyebrow of an incredulous skeptic, he evoked horrified shudders when he dramatically described hair-breadth escapes from the many ruffians sent to assassinate him. With humor, suspense, and thumping prose, Cassius Clay put on a good show. Some hearers declared that Clay's speeches made a "decided impression;" others—like the Birney men—derided him for the "*matter* of his discourse," as well as for his "miserable pettifogging *manner*."[17]

Such partisan comment, however, was but the common fruit of a campaign journey. His major effort still lay ahead, in New England and New York. From Niagara, the Clays took the train to Boston. As the car wheels clicked over the uneven rails, Cassius, considering the speech he would deliver, heard in their song a hymn to free labor. To New England, where he had first seen the vision of an industrial economy for his native Kentucky, he returned a decade later and renewed his allegiance to the American System and to its illustrious founder. Clay's speech in Boston's Tremont Temple demonstrated anew that his primary interest was the political defeat of slave-owners, because of their economic views. "Thus far," he asserted, "the pro-slavery power . . . has triumphed over the power of liberty and free labor." Southern slaveholders had monopolized the federal government, Cassius charged, and they had sabotaged the Whig program. "The system of internal improvements, as carried on by the General Government, and above all, the Tariff, have all been prostrated at the feet of the slave power." Now, with the tariff rates revised downward, "John C. Calhoun and his southern clique" sought new slave territory "to assist them in overthrowing the Tariff of Protection, and to reduce us once more to free trade and perpetual slavery." Again, in his campaign to New England abolitionists, Clay connected slavery with an economic principle. Earlier, addressing a New York audience, he had denounced the twin brothers, slavery and debt-repudiation. Now, in manufacturing New England, he decried the marriage of slavery with free trade. In each case he tailored his argument to his hearers; to neither group did he attack slavery by itself.[18]

In Boston, as elsewhere, Clay appealed to the self-interest of industrialists and of the laborers who depended upon manufactures for a livelihood. But as antislavery voters began to recognize, emancipation was not the ultimate object for which Cassius worked. As his campaign addresses emphasized, his purpose was to destroy the political domination of a small group of men—"John C. Calhoun and his southern clique"—

who defended slavery, but whom he disliked even more because they were the proponents of free trade, debt-repudiation, and limited internal improvements. In New England, as in Kentucky, he propounded an antislavery doctrine based upon economic considerations, and he was eagerly heard.

After filling his schedule in New England and addressing enthusiastic audiences, he moved down into New York, where the Thurlow Weed group, worried over the abolition vote, eagerly anticipated his presence among them. The state Whig leaders carefully planned his itinerary, hoping that he would convert Liberty Party adherents to Henry Clay. "I hope you will . . . give him such advice as you think useful touching his future movements," Congressman Washington Hunt told Weed, as they planned Cassius' campaign in New York.[19]

Despite the careful planning, however, Clay's mission was a failure. The group he had gone North to convert was not impressed by his arguments, and when he addressed them he evoked violent opposition. Although he tailored his arguments to fit the economic interests of northern financiers and industrialists, he refused to amend his antislavery reasoning to appease "higher law" abolitionists. He told them that slavery existed by local law, and that as long as the law existed, so might the condition. For such doctrines he made enemies among abolitionists who denied that law could sanction the evil of slavery. As they perceived his disavowal of their basic tenets, many of them became even more dissatisfied with the Whig Party. Cassius had been invited into New York to win them to Henry Clay, but because he did not speak from the motivation of religious zeal, Liberty Party advocates and Conscience Whigs began to criticize him and his candidate.

Although Lewis Tappan had previously admired Cassius, he now began to carp at him for endorsing the Whig candidate. "Henry Clay is a slaveholder—a duellist—a gambler—a profane swearer," Tappan piously declared. "How can a

Christian justify himself in voting for such a man?" Another Birney supporter rejected Clay's efforts to win over the abolitionists. "He was brought into the State doubtless for the express purpose of bringing us all over to the Whigs; but I presume he has done us but very little, if any damage. . . . I find that some of the Liberty men already consider him as a mere emissary of Henry."

The editor of a religious reform journal, the *Anti-Slavery Bugle,* gave his reason for rejecting Cassius: "C. M. Clay regards law as paramount to the rights of man. C. M. Clay said, 'That is property which the law makes property.' A more pro-slavery doctrine than this never fell from the lips of man. It virtually exalts legislative enactment above the government of God."

His refusal to temper his expressions to fit the prejudices of northern abolitionists made Cassius of little service as an emissary from his cousin Henry. Before the campaign had concluded, it was evident that the choice of Cassius M. Clay as representative of the candidate had been a mistake. But he was courageously consistent; he declared his opinion without regard for the consequences. Though he was ambitious, he refused to state a position in the North which he could not also support in Kentucky. In the existing state of national politics, then, he was a failure.[20]

Despite the criticisms of Liberty Party advocates, Cassius continued his efforts. As election day neared, he and Mary Jane turned homeward after three months of constant travel. The Clays were crossing the mountains to Wheeling while the voters were casting their ballots, and they did not learn the final result until they reached the foot of the mountain. There they saw a newly erected hickory pole, from which hung a skinned coon. By that mute symbol they knew that the Whigs, nicknamed Coons, had lost the election. New York, where Cassius had worked the hardest, gave its vote, by a narrow margin, to Polk. Clay's efforts in the state had not persuaded enough Birney supporters to vote for Henry Clay to enable the Kentuckian to carry the state.[21]

Though his candidate had lost, Cassius considered that the fight had only begun. He had become convinced that the sectional division over slavery would create a new political alignment. The Whig Party of the North, he said, in losing the moral issue of Texas on the slavery question, had lost everything. The South was solid in sacrificing all other issues to slavery, while the North wasted its votes on petty matters. Because the proslavery South had united, the free-labor party of the North must also unite, he declared. "Thus, and only thus, can the unholy and disastrous alliance between slavery, utter despotism, and so-called Democracy, be broken up." For if there was "anything worth preserving in republicanism," it would only be maintained, he grimly warned, by an "eternal and uncompromising war against the criminal usurpations of the slave power." The time had come, he announced, when the "friends of liberty" and the "craven slaves of despotism" must separate. To direct the new alignment of parties over the slavery issue, Cassius Clay prepared to renew his efforts as political propagandist in Kentucky.[22]

CHAPTER VI

CLAY DECLARES WAR

CASSIUS Clay's failure, and the subsequent defeat of the Whig Party, did not discourage him. Instead, the results of the election convinced him that the Whig organization was not an effective agent for unifying opposition to slavery; therefore, a new party should be formed. He advocated political union of the antislavery Whigs with the minority Liberty Party. "I look forward to the time not distant when the Whigs and Liberty Party will occupy the same ground," he told Liberty organizer Salmon P. Chase. "The Whigs number nearly one half of the nation, the Liberty men hold the balance of power." Such an organization, Clay contended, could win a national election without the assistance of southern voters. Ten years before the emergence of the national Republican Party, Clay saw the combination of forces which it later represented.

To help in bringing about the projected coalition, Clay planned a newspaper to advocate emancipation in Kentucky by political action. In January, 1845, soon after his return to Kentucky, he issued an "Address to the People of Kentucky," outlining his intentions. As usual, he introduced his program with a description of the economic effects which slavery had upon white labor by reducing the population and thus reducing the market for manufactured articles and food. Then he prescribed a political remedy. "*Let candidates be started in all the counties in favor of a convention,*" he urged, "*and run again and again till victory shall perch on the standard of the free.*" Still, he professed to be a moderate. "*Whether emancipation be remote or immediate,*" he said, "*regard must be had to the rights of owners, the habits of the old, and the general good feeling of the people.*" The emancipation he advocated would come gradually, after a long preparatory period in which he anticipated that most of the slaves would be sent out of the state. "To those who cry out forever, What shall be done with the free slaves? it will occur that upon this plan, no more will be left among us than we shall absolutely need." Clay recommended his scheme as one to satisfy all objections that might appear from the non-slaveholders.[1]

If Kentuckians would hear him, he would, he claimed, win their votes. His analysis of the state convinced him that he pursued a course which the majority should accept. "A space of three counties deep, lying along the Ohio River, contains a decided majority of the people of the state," he said. Because of their proximity to free states, their slaves might easily escape. "Soon, very soon, will they find themselves bearing all the evils of slavery, without any . . . remuneration." How long, Clay wanted to know, would they "tamely submit to this intolerable grievance?" He asserted that if slavery did not fall of its own weight, "they will vote it down, for they will have the *power*, and it will be to their *interest* to do so." [2]

The goal which he sought was constitutional emancipation of slaves, and he rejected any other solution. "I am opposed to depriving slave owners of their property by other than constitutional and legal means. I have no sympathy with those who would liberate slaves by any other means; and I have no connection with such people," he said. "I must, as a citizen, resist their efforts by force, if necessary." But he did not abandon his war upon slavery. "I am the avowed and uncompromising enemy of slavery, and shall never cease to use all constitutional, and honorable, and just means, to cause its extinction in Kentucky, and its reduction to its constitutional limits in the United States." [3]

Cassius announced that the time had come to establish an antislavery newspaper in Kentucky. There was a "large and respectable party, if not a majority of the people," he claimed, who desired to discuss emancipation and needed an outlet. "A press," he explained, "is only necessary to give concentrated effort and final success by free conference of opinion and untrammeled discussion." His paper, after the fashion of the day, would become the voice of a political effort.

Clay assured his fellow citizens that he would restrain his new publicity outlet. "Our press," he said, "will appeal *temperately* but *firmly* to the interests and the reason, not to the passions, of our people." Although he promised calm discussion, his temperament would lead him to heated invective. But from the beginning he expected the paper to lead a political organization. "We propose to act as a *State party*, not to unite with any party . . . ," he said, "expecting aid and encouragement from the lovers of liberty of all parties." And he proclaimed that no outsider would direct his efforts. "The times call for language plain, bold, and true—our cause is good—our press shall be *independent* or cease to exist." [4]

Determined to maintain his freedom, and having little faith in local law-enforcement, Cassius prepared his printing office. He leased a sturdy brick building on Mill Street in Lexington and called upon the adjutant of his militia regiment, Colonel Thomas Lewinski, to assist him in fortifying it. Lewinski, one of Clay's staunch supporters, was a Polish immigrant and a student of architecture and military engineering. Together, Lewinski and Clay lined the outside doors and window shutters with sheet iron to prevent burning. At the center of his defenses, Clay placed two brass four-pounder cannon upon a table facing the door and loaded them with shot and nails.

The entrance itself, protected by folding doors, was controlled by a chain which allowed only a small opening. Clay could fire his cannon through it, but attackers would have only a small target. Clay also stocked his fortress with firearms and with iron pikes modeled after Mexican Army lances. A platoon of loyal supporters completed the defenses. But should his office be successfully invaded, Clay had prepared an escape route through the roof. There a keg of powder and a match would provide final destruction of the office. One of his critics remarked that Clay acted "as though he were in an enemy country." Within a few months, Clay would agree that he was.[5]

His bristling defenses indicated that he anticipated trouble, and his editorial policy did nothing to avoid it. On June 3, 1845, *The True American* appeared, bearing a declaration of war upon slavery. He proclaimed that the issue was not whether "600,000" Kentuckians should "postpone their true prosperity" to the interests of thirty-one thousand slaveholders, but whether the majority would surrender its liberties to a "despotic and irresponsible minority." Clay suggested, in his party slogan, that his objective was twofold: constitutional emancipation to bring about "true prosperity," and a defense of the liberties of the whites. Cassius also made extremely violent statements about his opponents. For "Old Bob" Wickliffe, the man he used to symbolize the slave party, he spared no harsh epithet. "Old Man," Clay snarled, "remember Russell's Cave—and if you still thirst for bloodshed and violence, the same blade that repelled the assaults of assassins' sons, once more in self defence, is ready to drink of the blood of the hireling hordes of sycophants and outlaws of the assassin-sire of assassins."[6]

In girding for battle with slaveholders, Clay received encouragement from northern abolitionists who regarded his journalistic enterprise with delight and wished him well. "In a slave region a champion has arisen to battle against slavery," exulted the *Cincinnati Gazette*. At last, a "native of the soil, born and bred a slaveholder, disdaining to escape by flight, and brave enough to grasp the evil, and . . . to crush it in his own home, has flung his banner to the breeze. . . ." With less fervor, but with as much interest, Horace Greeley of the *New York Tribune* rejoiced to see that Clay exerted a "strong influence" in Kentucky. Busybody Lewis Tappan worried about Clay's defiant attitude. "He is bold and rather belligerent," Tappan said. "Whether he will be suffered to proceed or not is doubtful." But he concluded that if Clay could maintain his press, a "great victory" would be won. "Success will aid the Anti-S. cause in this country immensely," he told Clay. "May the God of the Oppressed sustain you. . . ." [7]

Among Clay's neighbors, however, there were some who denied that Cassius had any connection with the Deity. Lexington's *Observer and Reporter*, taking a neutral position, noted only that Clay's first issue had "relieved the somewhat monotonous and dull tone" of the local press by the "pungency of its assaults, and the severity with which it denounces." Another Kentucky commentator, more alarmed at *The True American*, declared that it was "thoroughly abolitionist," and, moreover, "insurrectionary in its character." Indeed, Clay had intended the first issue to be explosive. "Many complain of the rashness of my first number," he reported to Chase. "It was my deliberate judgment, not *passion only*. It was a necessary measure to *call up individuals* who were spreading the poison of mob violence, and make an issue *at once*. I have weathered the *crisis* I hope." Some of Clay's fellow citizens, however, were not convinced. "You are meaner than the autocrats of hell," they informed him, in a note written in blood. "The hemp is ready for your neck. Your life cannot be spared. Plenty thirst for your blood." Such advice would only inspire the defiant Cassius to more violent expletive and convince him that he was persecuted for defending constitutional rights.[8]

While admirers praised and critics threatened, Editor Clay busied himself with the burdensome trivia of newspaper operation. Although he had hired Frankfort journalist Thomas B. Stevenson to edit the paper, Stevenson subsequently refused to help him, and Clay was left with the full responsibility himself. His ambitious journalistic program made it a time-consuming project. *The True American* was a full-sized newspaper of four pages, each eight columns wide and nearly a yard long. Scissors and paste pot occupied prominent positions at Clay's ponderous roll-top desk, for much of the paper was given over to editorial copy from fellow journalists. Approximately two columns on the back page contained advertising of the usual ante bellum wares: patent medicines, machinery, books, business houses, and the like. But despite the copious copying and the advertising, editing the paper was a full-time job which Clay soon came to detest. He did not like to be bound by a weekly deadline; he was a man of action as well as of words, and he preferred excitement and activity to the dreary, humdrum task at the desk.

Clay's task was not complicated, however, by a search for fresh material. To those who had read his essays since 1841, the editorial columns of *The True American* offered little that was new; he simply repeated what he had already said on numerous occasions. The paper was primarily an instrument for antislavery agitation and for the formation of a new political party to act against slave-owners. Once more Clay documented the pernicious effects of slavery by comparing the free with the slave states, and always the free states won. "If a single State only illustrated this contrast," he said, "there might still be room for argument. But here are twenty-six States covering a continent . . . and yet thirteen times has this struggle for ascendency between liberty and slavery taken place . . . and thirteen times has liberty borne off the palm." The reason for that situation, he argued, was as "shallow and transparent" as the result. There were three millions of slaves in the South who performed, he alleged, about one-half of the effective work of the same number of whites in the North, because they lacked the "stimulus of self-interest."

Clay also reiterated his condemnation of slavery for its obstruction to industrial development. He pointed out that all the necessities and luxuries consumed in the South had to be imported, and remarked that of course double freights had to be paid by her. Manufacturers could not thrive in a slave area, however, because the slave population did not consume enough to make it profitable. "Lawyers, merchants, mechanics, laborers, who are your consumers, Robert Wickliffe's 200 slaves?" he asked. "How many clients do you find, how many goods do you sell, how many hats, coats, saddles, and trunks do you make for these 200 slaves?" Would Wickliffe, Clay wanted to know, purchase as much for his Negroes as two hundred free white laborers would buy for themselves? "We stand for the whites; Mr. Wickliffe for the slaves," he added.[9]

He attempted to drive a wedge between slaveholders and non-slaveholders, for only by breaking the solid front of the whites could he effect a revolution at the polls. He urged non-slaveholders to exercise their ballots to defeat the system which harmed them. "Yes, thank God, we can yet vote! . . . Let us but speak the word, and *slavery shall die!*" Constantly he encouraged his readers to act, regardless of the costs. "You will be assaulted and shut in on all sides, traduced in your character, injured in your persons, in your business, and in your families," he warned, from his own experience. "Never fear, brave hearts," he counselled, "oat meal can be had at twenty cents per bushel, they can't starve us yet. . . ." Clay would discover, however, that few Kentuckians cared to dine on oatmeal or shared his zeal for martyrdom.[10]

It was to the laboring class that he made his most earnest appeal. Non-slaveholders, who shared none of the benefits of slavery, paid its cost, he said, in reduced wages. "Thus every laborer in Kentucky is injured by the one hundred and eighty thousand slaves," he concluded, "as if the same number of Irishmen, Dutchmen, or Englishmen should come in here and agree to work as the convicts or the slaves do, without wages." For that reason he urged workingmen to vote against slavery: "Shall we any longer support it, by our countenance, or our votes?" Always, whenever he considered the solution for the evil, Clay came back to the ballot box; he looked to political action to bring about a change.[11]

While he devoted much of his editorial effort to a continuing defense of the Kentucky System and to a political solution to the encroachments of slavery, he also lightened the pages of his journal with lusty humor. Solemnly he compared divorce statistics in slave and free states and discovered that divorce was more prevalent in slave states. He concluded that broken homes resulted—like all other human ills—from slavery. "Put away your slaves," Cassius exhorted southern matrons. "Make your own beds, sweep your own rooms, and wash your own clothes; throw away corsets, and nature will form your bustles.—then you will have full chests . . . and *no divorces*." Clay caricatured the idle, overstuffed, slave-pampered women as "forked radishes." In another issue he lampooned a "silly girl" who heard that a small waist was becoming. She wrapped herself with "silk cord and canvas, till a man would sooner put his arms around a lamp-post, than one of these unpliant, mummy-wrapt sticks." [12]

But along with spicy dashes of humor, the vinegar of a bitter hatred also flavored the repast which Clay offered his readers. Repeatedly he challenged the slave party. "We hurl back indignant defiance against the cowardly outlaws," he exclaimed. "We can die, but cannot be enslaved." Over and over again, in heavily italicized phrases which emphasized his fervor, he pronounced the doom of slavery. "The slaveholders and their *sycophants* will find," he rasped, "that the *free white laborers of this land, composing four-fifths of the popu-*

lation, . . . are not slaves. Slavery is doomed—it must die!—*the first act of violence in its cause, will hasten its fate!*" [13]

Slave-owners returned Clay's hatred, with an antagonism that stemmed from a source deeper than annoyance at a few intemperate sentences. Slowly, *The True American* was beginning to make its influence felt in Kentucky. For that reason, rather than because of Clay's outspoken venom, they feared Clay and his paper. Two months after the paper first appeared, the subscription list had grown from three hundred to seven hundred in Kentucky, and from seventeen hundred to twenty-seven hundred in other states. Clay generously estimated that twenty persons in the state perused each copy, making fourteen thousand readers in Kentucky. The record of the paper's growth convinced the editor that, as he later said, "the principles and tone of my press were taking a powerful hold upon the minds and affections of the people." Perhaps, had he not been so effective, he would have been unharmed.[14]

The journal proved, however, to be a spark that ignited fires in other parts of the state. A Methodist weekly in Georgetown, Kentucky, the *Christian Intelligencer*, proclaimed its adherence to Clay's program. In Louisville, where there was a strong commercial community, Clay's friends planned to begin another emancipation newspaper. In the same city, a candidate for the legislature announced his approval of constitutional emancipation and offered himself as a representative of the Emancipation Party. In the Green River section—"the most pro-slavery part of the state"—a Democratic paper copied some of Clay's articles describing the competition which slavery offered the white laborers and "seemed ready to wage a common war." Observing all these signs, Cassius declared that his dream of an independent emancipation party was nearing reality.

On the basis of the apparent strength of emancipation sentiment, he appealed for the support of other politicians who would gamble upon future victories. "Where is the man . . . ," he asked, "who will sacrifice present power, to the contingency of hereafter rising with the swelling tide of freedom?" Cassius Clay had already decided upon such a sacrifice, and he eagerly waited for the "swelling tide" to develop.[15]

Even without the assistance of other politicians, however, Cassius proclaimed that the Emancipation Party had at last been established. In mid-July he exulted, "The seeds of an independent party is [*sic*] planted—a party of slow but sure growth, but of certain success and lasting power." For the first time in the state's history, he boasted, a political party had organized "for the overthrow of slavery in a legal way." Candidates were in the field, newspaper circulation was growing, and plans were made for a state convention of the "friends of emancipation" to be held July 4, 1846. "And the great mass of laborers, who are not habitual readers of newspapers," Cassius exclaimed, "began to hear, to consider, and to learn their rights, and were preparing to maintain them." [16] But Clay's hopes for immediate success proved premature. In August, only a month after he had gloried in his triumph, he experienced the full force of the slave-owners' wrath.

CHAPTER VII

THE EIGHTEENTH OF AUGUST

IN JULY, 1845, when he considered himself near success, Cassius Clay fell ill. An epidemic of typhoid fever hit Lexington, and he contracted the disease. Although he should have delegated his editorial duties to his assistants during his extended illness, he attempted to direct the press from his bed. In doing so he precipitated the most serious crisis of his career: he approved an article and penned an editorial which aroused the slave-owners to action against him.

The pretext which they employed to attack the paper was the publication of a bit of impassioned prose hardly more inflammatory than much he had already written. The article, voluntarily submitted by a correspondent—a slaveholder, Cassius insisted—was entitled "What is to become of the slaves in the United States?" and suggested radical changes in the treatment of freedmen. Its author contended that slavery did not pay a profit and would eventually disappear. Negroes would then become eligible for citizenship. The columnist urged political equality for Negroes already free, to prepare the way for eventual liberation of all slaves. The essay contained phrases which easily lent themselves to misunderstandings, and an alert editor would have rejected it, particularly for publication in a slave state.[1]

Unfortunately, however, Cassius was wracked by his illness

and did not observe the danger signs. Indeed, he later protested that he had not even read the article prior to its publication. But in his first issue he had committed himself to publish divergent opinions in his paper. He had no other reason to accept the article, for it contributed nothing to his purposes. He had long advocated a plan which would obviate the problem of the freed Negro, at least in Kentucky. Furthermore, he had frequently demonstrated that he had no love for Negroes, slave or free, and cared nothing about granting them equality. The article was, therefore, out of tune with his own program and prejudices, and Clay erred in accepting it for publication.

Not only did he print it, however, but he accompanied it with an editorial which contained a dangerously ambiguous paragraph. His hatred for the aristocratic slaveholders he reiterated in words which seemed to invoke armed revolution. "But remember, you who dwell in marble palaces, that there are strong arms and fiery hearts and iron pikes in the streets, and panes of glass only between them and the silver plate on the board, and the smooth-skinned woman on the ottoman," Clay exclaimed. "When you have mocked at virtue, denied the agency of God in the affairs of men, and made rapine your honied faith, tremble! for the day of retribution is at hand, and the masses will be avenged." Coming from a man known to have iron pikes in his fortress, the editorial aroused community leaders to action. They seized upon the resulting public excitement to take measures against Clay's paper.[2]

On Thursday afternoon, August 14, two days after the article appeared, several influential citizens called a meeting to plan their action, before the excitement should abate. Clay, bedridden in his North Limestone Street home, heard of the session shortly before it convened. Despite the remonstrances of his family, he arose, dressed, and rode to the courthouse in his buggy. When he entered the hall, pale and trembling, he found more than twenty proslavery leaders plotting against his journal. Too weak to sit up, Cassius reclined on one of the benches and heard one after another of the men denounce *The True American* as insurrectionary and as an intolerable public nuisance. Occasionally he arose and attempted to respond, but he was refused a hearing. When he saw that his continued entreaties were to no avail, he left the hall and returned to his bed.[3]

When he got home he prepared an account of the whole affair to be issued the next day as a *True American* extra. Then he made plans for the defense of his printing office, warned his cohort of trusted friends of the threatening crisis, wrote his will, and sent a camp bed to the office. Grimly he determined to resist whatever action the citizens' meeting should propose.[4]

He had just completed these tasks when a delegation from the "respectable citizens of the City of Lexington" arrived with a message. Declaring that *The True American* was "dangerous to the safety of our homes and families," the community spokesmen requested Cassius to discontinue its publication. "Your own safety, as well as the repose and peace of the community," the group warned him, "are involved in your answer." They branded Clay a rebel and a man dangerous to the public welfare. Clay's opponents did not have to answer his arguments when they could arouse fear among non-slaveholders by quoting his own words. And fear was on their side —fear of the consequences of emancipation. Many non-slaveholding citizens were sincerely convinced of the necessity of maintaining slavery as the simplest solution to the race problem.

The request that he cease his journalistic activities infuriated Clay, and he delivered a fiery response. He pointed out that he was helpless from the fever and charged that the attack upon him was therefore dishonorable. The citizens' meeting was an extralegal affair to which he could give no recognition. "I deny their power and I defy their action," he answered. Though he was weakened by illness, his spirit was undaunted. "Your advice with regard to my personal safety is worthy of the source whence it emanated, and meets the same contempt from me which the purposes of your mission excite," he snarled from his bed. "Go tell your secret conclave of cowardly assassins that C. M. Clay knows his rights and how to defend them." [5]

While Cassius defiantly answered his enemies, relatives urged him to surrender to the determined community leaders. His mother, however, although she understood his "quick and hasty temper," and had feared he would fall into trouble, insisted that he follow his own conscience. "Cassius, don't give up anything you think it your duty to defend," she said. "If you prefer death to dishonor, so do I." Her son and his wife, equally determined, resolutely faced the crisis.

Clay kept his press in operation throughout the weekend, publishing a series of handbills which explained his side of the argument. He also sought to allay the hostility which the slave-owners were building up against him. He dictated all of the statements to an amanuensis while his hands and his head were continually bathed in cold water to reduce his raging fever. The day following the committee meeting at the court-house, Clay published an extra issue of *The True American*. He said that the committee leaders were politically opposed to him and attacked his paper for that reason rather than for what he had printed. They had referred to him as a revolutionary who plotted a slave insurrection. In a state where there were six whites to one slave that fear was, he said, ridiculous. His appeal, he repeated, had been to the state's six hundred thousand free whites, rather than to slaves. He called upon non-slaveholders to consider his fight their own struggle for freedom. He asked them where they would stand when the "battle for liberty and slavery" would begin. "If you stand by me like men," he said, "our country shall yet be free." He still professed to defend the rights of the majority and was determined to fight for his freedom. Once more, however, he had provided his enemies with more evidence that he contemplated armed rebellion.[6]

On Saturday, August 16, in an effort to dispel the charge that he was a revolutionary, he published another manifesto outlining his plan of emancipation. His enemies, he said, insisted upon calling him an " 'abolitionist,' a name full of unknown and strange terrors and crimes, to the mass of our people. . . ." Again Cassius declared that he had no connection with the abolitionist movement, and he repeated his desire for gradual and legal emancipation. He explained that he favored the election of a convention to amend the state constitution, with the proposition that every female slave born after a stipulated date (such as 1860, or 1900) should become free at the age of twenty-one. Because the common law provided that the status of the mother determined that of the offspring, eventually all slaves would become free. Such a gradual scheme would allow years to adjust to liberation. Clay repeated his prediction that, as the effective date approached, many owners would sell their slaves out of the state to avoid the loss.

Cassius also repudiated the article which had been the source of the current unrest. He asserted what he had always maintained: that he opposed giving political equality to freed slaves until they could exercise it with discretion. Having removed the basis for the agitation against him, he made his first conciliatory offer. "I am willing to take warning from friends or enemies for the future conduct of my paper," he said, "and while I am ready to restrict myself in the latitude of discussion of the question, I never will voluntarily abandon a right or yield a principle." There was a hint of his fiery temper in the declaration, but he was a more chastened man than he had been the day before.[7]

And with the passage of another day, Cassius retreated still further from his original defiant position. On Sunday he circulated another broadside addressed "To the People." He reminded his fellow citizens that he had been ill for thirty-three days, with his brain "almost incessantly" affected by the fever. Because of his illness, he lamented, he was unable to "pull a trigger or wield a pen," and in that helpless condition he was attacked. His foes had misrepresented his publications in order to arouse public opinion against him, he protested, and they had incorrectly associated him with northern abolitionists. "I utterly deny that I have any political association with them," he said.

Sober consideration of his own weakness led Cassius to surrender yet another of his defenses. Sunday evening he ordered the removal of the muskets and "other deadly weapons" from his fortress-office and decided not to defend the building. Had he continued his original determination to fight the public action, he would have given a good account of himself from within the office, but in his weakened condition he would eventually have been taken. The effects of a lingering case of typhoid fever saved him from becoming a second Lovejoy.[8]

As darkness fell over the community on the eve of the mass meeting which had been scheduled for Monday, August 18, many people fancied that they detected signs of insubordination among the slaves. During the night, patrols of armed men kept watch in the hushed, dark streets. Five blocks out on Limestone Street, in his home, lay the crusader whose impetuous activities had caused the excitement. By the flickering yellow lamplight which played upon his wan face, Cassius dictated his final message before the Monday meeting. In a handbill to be given to the citizens who attended the meeting, he finally surrendered everything but his right to publish a newspaper. He admitted that the editorial was more inflammatory than he had intended, but he begged to be excused because of his illness. He promised that in the future he would admit no article to his paper for which the public could not hold him accountable. "This, you perceive, will very much narrow the ground," he said, "for my plan of emancipation . . . is of the most gradual character."

In addition, he made another conciliatory statement. Until he had recovered sufficiently to supervise the press, no further comment on the slavery question would appear. He would continue to issue the paper but would temporarily cease his agitation for an emancipation party. In conclusion, he announced that his office and his dwelling were defended only by the laws of the land, adding, "And of those laws the citizens are the sole guardians." From his original defiant determination to defend his rights, Cassius Clay had backed down until he had surrendered not only his office and his home, but also his activities for political publicity.[9]

Clay's last-minute promises did not stay the proslavery leaders. At eleven o'clock in the morning, on Monday, August 18, a gathering of citizens convened at the Fayette County courthouse. Because the courtroom was too small, the meeting was held on the lawn behind the building. There, Lexington's citizen-body, more than two thousand strong, heard orator Thomas F. Marshall arraign Clay as a dangerous insurrectionary. Marshall, a nephew of the great chief justice of the Supreme Court, John Marshall, privately educated among his Virginia relatives, was an able lawyer. The indictment he brought against Cassius Clay was a polished legal brief. It was an appeal to a judge and a jury, in this case, the assembled convocation.[10]

In his prosecution Marshall enumerated Clay's heretical opinions. Cassius had advocated the "Abolition of Slavery in the District of Columbia . . . the exclusion of the three-fifths of the slave population in the apportionment of representation by a change in the constitution . . . the exclusion of Texas from the Union . . . the enlisting of the whole force of the non slave-holders in Kentucky against slave property, . . . thus forcing a change in the constitution of the State." These, Marshall charged, "were among the means and instruments relied upon by him for effecting the entire abolition of slavery in America." Clay's work among the non-slaveholders, however ineffectual and however legal, was condemned as a crime for which slaveholders demanded retribution. "That this infatuated man believed that the non-slaveholders of Kentucky would feel and act as a party against the tenure of slavery, and that through them he expected to change the constitution of Kentucky, and finally overthrow the institution," Marshall exclaimed, "is evident from one of his letters to the Tribune." Clay's crime lay not in uttering an unguarded sentiment or in appearing to contemplate force. To Kentucky slave-owners, it was enough that he sought to construct a law-abiding political organization in opposition to their interests.

Having thus condemned Clay for his political activities against slavery, Marshall came to the topic of the day. The mass meeting had been called, he told the assembled audience, to discuss measures for the suppression of the paper. As that was palpably an unconstitutional action, Marshall's task was to render it acceptable. The long hours spent perusing the casuistry of legal terminology now paid dividends. Marshall admitted that freedom of the press and freedom of discussion were basic human liberties. But, he insisted, Clay had used that freedom to plot domestic violence and had, therefore, forfeited his liberties. Clay had begun publication with the anticipation of employing force to gain his ends. He had fortified his printing office as though he were in an enemy country and had stocked it with "mines of gunpower, stands of muskets and pieces of cannon." Either Clay was a madman, the orator reasoned, or he planned a civil war in which he expected non-slaveholders and slaves to join him, to fight for abolition. Sensible people, Marshall assured the Lexingtonians, would not allow such an incendiary threat to remain among them. "An Abolition paper in a slave state," Marshall declared, "is a nuisance of the most formidable character . . . a blazing brand in the hand of an incendiary or madman, which may scatter ruin, conflagration, revolution, crime unnameable, over everything dear in domestic life." Who should say, he continued, that the "safety of a single individual is more important in the eye of the law than that of a whole people?" Adroitly Marshall argued his case against Clay, using the age-old controversy of freedom versus order. Marshall's argument implied that there could be no discussion of slavery in a slave state. Cassius Clay had sought to exercise a power which slave-owners regarded as non-existent.

Thomas F. Marshall branded Cassius M. Clay a trespasser in Lexington, an invader who intended to destroy the foundations of society. The committee of citizens had courteously requested Clay to cease his journalistic efforts, he said, and the editor had arrogantly refused. Now, Marshall shouted, the sovereign people would remove him by force. He concluded his brief with suggestions for enacting that verdict. He asked the citizens to ship *The True American*—presses and apparatus—beyond the state line. Should there be any resistance to their action, they should force the fortress-office "at all hazards, and destroy the nuisance." As their agents to perform the task, Marshall recommended the appointment of sixty men to take possession of Clay's property and to supervise its shipment.[11]

Without a single dissenting vote, the convocation adopted Marshall's resolutions. Chairman Waller Bullock called off the names of the Committee of Sixty and appointed George W. Johnson as its chairman. In accordance with its instructions, the committee proceeded to the Mill Street office of *The True American*. As he had promised, Cassius Clay made no attempt to defend his property. Earlier in the day, he had surrendered the key after receiving a court injunction upon

Cassius M. Clay as a young man. Courtesy of Warfield S. Bennett.

Cassius M. Clay, "Champion of Liberty." From a print by Currier.

Republican candidates for nomination in 1860.
From Harper's Weekly. *Courtesy of the New York Public Library.*

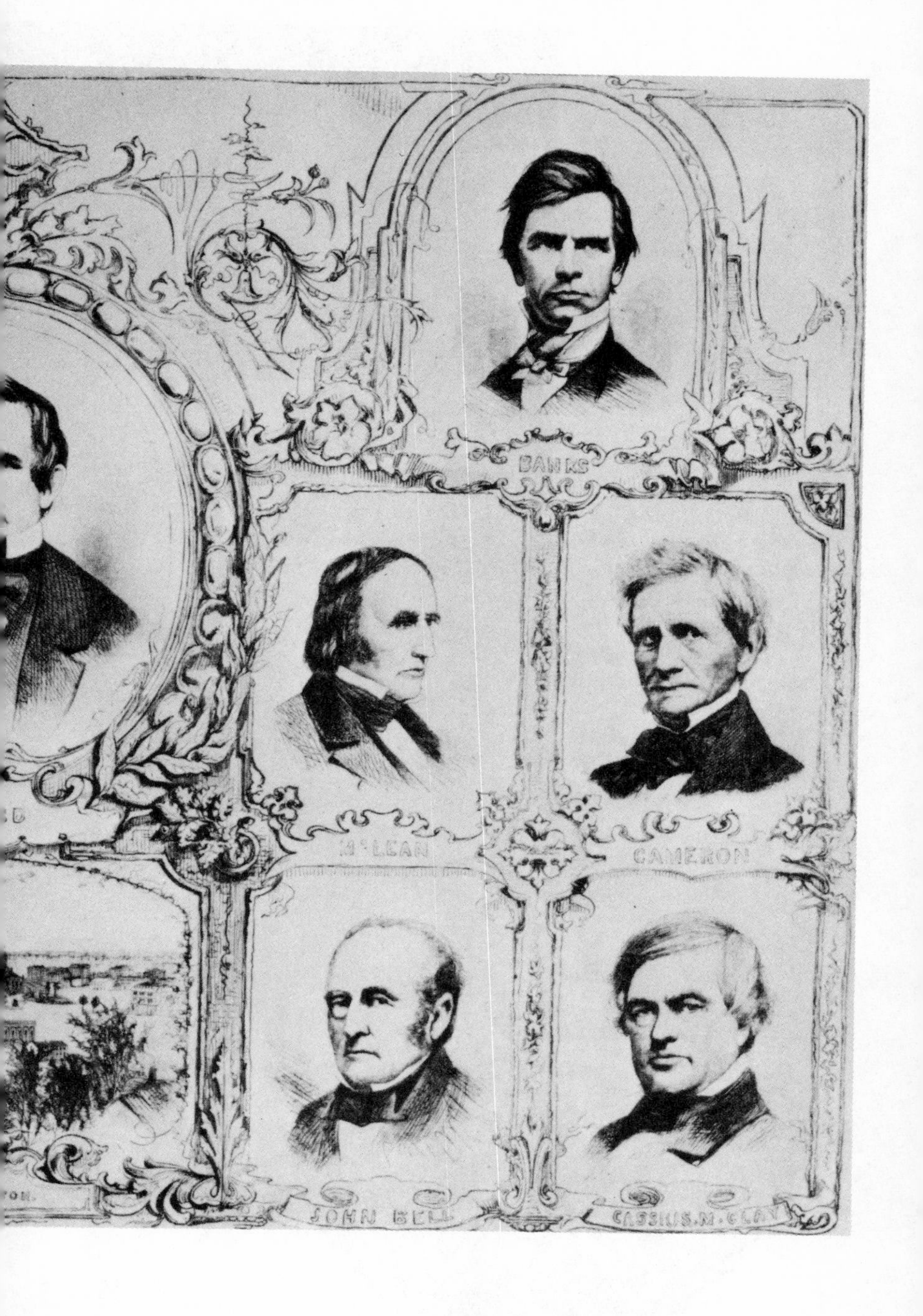
BANKS
McLEAN
CAMERON
JOHN BELL
CASSIUS.M.CLAY

Mary Jane Clay
as a young woman.
From a portrait by Healy.
Courtesy of
Miss Helen S. Bennett.

Mary Jane Clay
in Russian Court Dre
Courtesy of
Miss Helen S. Bennett

his property. James Logue, mayor of Lexington, was waiting at the door to serve notice that the committee acted in opposition to law, but that the city authorities could offer no forcible resistance.[12]

Inside the shuttered building, the committee set about its task. James B. Clay, son of Henry Clay, served as committee secretary and kept the records of their operation. They sent the desk, containing the private papers of the editor, to his home. Then they summoned master printers to direct the packaging of all type, presses, and other articles belonging to the paper. With great care, guided by professional advisers, the committeemen crated the expensive fixtures. Before the day had ended, their work was complete. They delivered the boxes to the railway station for delivery to a commission firm in Cincinnati.

Lexington citizens boasted of their self-control in dealing with an antislavery organ. A Lexington editor recalled previous occasions in other communities in which similar action had been taken. His city alone, he said, had exhibited the rare spectacle of a body of citizens, aroused over an incendiary press, yet so well controlled as to "accomplish their purpose without the slightest damage to property or the effusion of a drop of blood." [13]

While the citizens of Lexington boasted of their moderation, Cassius Clay had attained martyrdom in the eyes of the northern abolitionists. A Cincinnati writer sang a hymn of praise to Clay: "He braved a tyrant power with a courage great as the occasion," and if he had temporarily fallen it would only be to rise again, "as giants rise when refreshed by sleep." Another Ohioan rejoiced that Clay had "grappled the monster in his den—sacrificed his property, and offered his life in the defence of a cause he knew to be just." An abolition editor called the incident an attempted murder. "If Clay dies," he said, "he will be the victim of slave-holding mobocrats." And an antislavery orator declared that Clay was the agent of the Almighty against slavery. He said that Cassius had received his authority "from the framer of the highest constitution and laws known to man, by the commands of the living and eternal God. . . ." [14]

Other northern antislavery spokesmen received Clay's "sacrifice" with more caution. The *Anti-Slavery Bugle*, organ of religious abolitionists, criticized Clay's principles while approving his stand. "It is pretty well understood that we do not regard Cassius M. Clay as an abolitionist occupying the true position," its editor remarked, "but as one who opposes the institution of slavery in a manner and by means of which we utterly disprove;" but he went on to commend Clay "as an honest foe to that accursed system." William Lloyd Garrison, self-appointed defender of the faith, also condemned Clay's program while exulting over his courage. Cassius had gained nothing by remaining aloof from immediatism, Garrison claimed. Cassius Clay was a southerner, and not an outsider; he was no fanatic, but a "talented, high-minded, independent" person; he had explicitly rejected violent or immediate emancipation and had repeatedly denied any association with abolitionists. Yet none of those facts had saved him from the violence of the slave-owners. "What has he gained by refusing to occupy the ground of northern abolitionists?" Garrison demanded, and answered, "Nothing. What has he not lost?" But the *Liberator's* fiery editor rejoiced because Clay's experience would make "thousands of converts to the antislavery movement," and would "confound the enemies of freedom." In New York, Lewis Tappan sternly denounced Clay's efforts to maintain a middle ground. "If a man gives up, or does not embrace *immediatism* I think his anti-slavery essays will do little good," Tappan concluded.[15]

While the debate over his actions continued, Cassius recuperated from his illness. Soon after the excitement of the eighteenth of August, he left Lexington and went to Estill

Springs to "rusticate and cool off," as one observer commented. His sickness had proved to be his salvation. Had he been in health he might have felt obliged to defend his office against determined proslavery hostility. But thanks to the disease, he maintained his reputation as a fighter without pitting his strength against an overwhelming force. In addition to giving him an excuse for evading a fight, the illness provided him with an effective argument. He would declare that his enemies had craftily awaited his indisposition to attack him. "I believe now, as ever," he said in 1848, "that had I not fallen sick, I would never have been mobbed." [16]

Though Clay had gained another argument to use against his foes, the success of the proslavery party in enlisting mass approval of its action indicated that he had suffered a major defeat. The program he advocated was legal, gradual, and constitutional; the end he sought was economic prosperity for the white man, rather than liberty or equality for the Negro. But though he preached a peaceful plan, his enemies were able to distort it by referring to his warlike language and his fortified office.

The real reason for the suppression of *The True American*, however, lay deeper than Cassius Clay's belligerent temperament. His efforts to institute an emancipation party in a slave state angered and frightened the dominant slaveholders. Lanky Tom Marshall had listed Clay's political efforts as "crimes" against the community, and though he ridiculed him, he betrayed the deep fear which obsessed his compatriots. An emancipation newspaper in a slave state was a nuisance, he declared, implying that however it operated it would not be tolerated. Clay's political ambitions had clashed with the deep-seated fears of the predominant slave aristocracy, who trembled at the thought of his possible success. Many others supported the suppression of Clay because of their fear of his success—fear of still another kind. The specter of the freed Negro, perhaps competing for white men's jobs, moving in great numbers into the towns, was an obstacle Cassius never overcame. For such reasons, the citizens of Lexington would not allow his press to remain among them.

Although Clay's plans had suffered a setback, he was not discouraged. As soon as he regained his strength he renewed his efforts for a political victory over his opponents. Although they had discredited his newspaper and perverted his program, Cassius Clay was not ready to surrender.

CHAPTER VIII

"THE MOB WILL NOT STOP ME"

AFTER overcoming the effects of his lingering illness, Clay immediately planned the revival of *The True American*. In the face of determined opposition he revealed the courage and the perseverance which characterized his career. He had sustained a defeat, but he was undismayed. "The mob will not stop me," he told his New England subscription agent. "*Somewhere, I will go on soon.*" He admitted that slaveholders had expelled his press, but, he added defiantly, "there are not men enough in *Kentucky* to '*drive*' us out of the state." Boldly and unhesitatingly Cassius renewed his war upon the slave party.[1]

His first effort was to seek legal redress for his loss. On September 18, exactly one month after the suppression, the Committee of Sixty faced trial for their part in depriving Clay of his property. After hearing a full review of the circumstances, the jury rendered a verdict of not guilty to the charge of committing a riot. When he failed to win his case, Clay contented himself with embarrassing the committee members by refusing to call for his press at Cincinnati. They had to continue paying rental and storage fees much longer than they had expected, for if they did not, and the goods were seized, Cassius could charge them with theft.[2]

On October 8, about a week after the trial closed, Cassius

resumed his campaign against slave-owners by reviving *The True American.* Though the paper bore a Lexington date line, it was printed in Cincinnati. As the months passed, weekly editions of the resurrected paper appeared regularly, but it became increasingly evident that the suppression had been a defeat for Cassius Clay. No longer did he boast of a growing emancipation party. Instead, he watched as the slaveholders organized opposition to him. All over Kentucky, they held public meetings for the purpose of censuring him. The editor of the *Observer and Reporter* declared that there were more such meetings than he had room to describe, and he exulted that "all approve, while not a murmur of discontent is heard." [3]

The self-appointed spokesmen for Kentucky public opinion had united upon the same charge against Clay and his press. Invariably they denounced him as an irresponsible firebrand who endangered the community. The common argument was clearly expressed in the *Observer and Reporter*. "To put such a lever as the Press into the hands of such a man as C. M. Clay, heedless, reckless, impetuous, ultra, and revolutionary," the editor said, "is almost like putting a torch into the hands of an incendiary." None of Clay's detractors answered his arguments and declarations, but all of them used his own belligerence against him.[4]

That same trait impelled Cassius to maintain his position at all costs. Valiantly he forced himself to remain at his desk and continue his paper. More and more he felt the burden of the weekly deadline, and he longed to relinquish the confining duties of the editor's chair. "The conducting of a newspaper is neither suited to our early habits, our tastes, nor our necessities," he lamented. Nevertheless, he doggedly continued his efforts to inform non-slaveholders of their rights. No longer was he certain that it would be a simple matter to build an emancipation party. His ambition was too strong, however, to allow him to give up because the task was difficult.[5]

After re-establishing the paper, Clay's first concern was for its subscription list. Because he aimed his editorial matter at non-slaveholders in a slave state, he had to reach large numbers of them in order to effect his program. To expand his circulation among the group he wanted to influence, Clay offered his paper to them at one-half the regular rate. He explained that he had established *The True American* to inform them of the oppressive nature of slavery, but the slaveholders "arose in arms" and suppressed it. "Because they saw well enough," he said, "that, *if you once learned your rights, Slavery*, as you had the power, being about five freemen to one tyrant, *would be destroyed!*" The reduction in the subscription rate, he generously admitted, would not profit him, except that if he succeeded he would "partake in the common welfare and happiness of the people." [6]

To further his cause, Cassius added a new element to his antislavery arguments. He discovered that his repudiation of the religious basis of the antislavery crusade had estranged possible followers. After the suppression of his paper, he spoke more of moral commands and religious conviction than he had done before. In December, 1845, he published a brochure *To all the Followers of Christ in the American Union*, asking their support in his struggle for freedom. The columns of *The True American* also reflected a new moralistic approach to the slave issue. He still did not appeal to the religious sensibilities of slave-owners, but he attempted to bestir the consciences of non-slaveholders to vote against the evil. He also continued to repudiate the organized religion he saw around him. "The church . . . is *slimy* and *false*," he said; "there's no soul in it with few honorable exceptions." He especially attacked the religious defense of slavery. "We despise your slaveholding religion," he told his fellow citizens. Not one of the seven churches near his home, "which continually annoy us with their everlasting bell-ringing," had stood by him in the recent excitement. "As we stood for the rights of man, the liberties of our country, and the purity of Christianity, they were silent." To attract the moralists whom he had previously neglected, Cassius attempted to distinguish between a "slave-holding religion," and "pure Christianity," which was, by his definition, at war with slavery. "We oppose slavery, not because it obstructs us in the race for life," he said, explaining the new addition to his repertory of antislavery arguments, "for it does not, seeing we had the vantage ground by birth; but because it is at war with nature and the laws of nature's God." [7]

Along with the addition of a moralistic dart to his quiver of arguments, Cassius did not lose sight of his interest in the Kentucky System. Within a month of the paper's reappearance he had repeated the description of its operation, painting its advantages in idealistic phrases. Under his plan of emancipation, as he had said before, he anticipated that most of the blacks would be sold out of the state, "and thus relieve our people from their imaginary difficulties, of a large free population." All the slaves sold would provide capital, he promised, "ready to be invested in manufactures," which would entice white laborers and also "men of capital" into the state. "Thus would the towns begin to grow once more; . . . and home markets be secured for the productions of the soil. . . ." Laborers would find work and would then consume the products of the towns and of the farms. "There would be no more fears of insurrection, civil war, and unknown disaster," he promised. Prosperity, peace, progress—these were the delightful vistas which Cassius Clay's energetic imagination depicted in support of his plan to introduce a balanced economy into the state.

Although he hurled moral and economic denunciations at slavery, it was through the ballot that he sought reform. "*I can vote*—nail all such maxims to the masthead . . . ," he ordered. "You are in the majority. Assert what is right, and do it, and the day is yours." Nor did he cease his efforts to encourage office-seekers to represent emancipation. "*We ask the five hundred thousand white non-slaveholders to make these tools of slaveholders* [the state officials] . . . meet the doom of traitors!" Clay demanded. "*Let us see if we can't find some other men than they, to represent* FREEMEN." He asked what kept Kentucky from such a revolution, and responded, "Nothing but the want of unity and energy. Give us these," he implored, "and let the voters of mountain and lowland speak out for freedom. . . ." [8]

Clay appealed for support in all sections of the state, but he was beginning to suspect that he would receive a more sympathetic hearing among the mountaineers in potentially industrial areas than among the inhabitants of the lowlands. He charged that the tobacco interests of the state were his fiercest enemies, and that they led the fight against him. But his right to discuss emancipation was, he declared, "most ably sustained by the mountains where few slaves exist." After the defeat in the Bluegrass, as he recalled many years later, "I turned my eyes toward the mountains eastward, where few slaves were held." He rejoiced to consider the support of hill folk. "It proves," he boasted, "that the *true issue* begins to be understood, and that we, the *non-slave-holders* of this State, are destined to overthrow slavery." He pointed out that there would always be a border country between slave and free sections, "and no state, except Louisiana, is without its mountains and its mountain men." And in no such area, he continued, "can slavery find long a resting place." The spirit of freedom which permeated the air along the border, and which filled the smoky breezes of the hills, affected both master and man, Clay said, and would sweep away the institution.[9]

Cassius Clay had recognized a fundamental fact about the diversified ante bellum South; in the southern mountains lived men who owned land but who did not own slaves. Among Clay's contemporaries it was widely believed that southern landowners were also slave-owners, but there was one great geographic area in which the generality did not hold. Mountaineers, confronted with a topography unsuited for the culture of uninterrupted hundreds or thousands of acres of land, generally did not command the income of the plantation. In that area slavery did not pay. Hill folk had little in common with the men who dominated the lowlands and who often decided policies for the people in the mountains as well.

Ever since the planting of the American colonies there had been an internal conflict between the gentry of the Tidewater —the narrow strip of fertile coastline extending back to the fall line of the rivers—and the equally proud residents of the back-country hills. Their interests conflicted; and in most colonies, and later in the states, the political machinery operated for the benefit of the coastal gentry rather than for the hill folk. In Kentucky, the geography was reversed, but the struggle was similar. The hills were in the east, on the slope of the Appalachian range, and the lowland plain was in the west, but there was an internal battle between men of the two sections. The Inland Tidewater formed by the Ohio and Kentucky River valleys, with its rich soil, favorable climate, and slave-owning gentry who managed the hemp and tobacco farms, resembled the coastal plains. But in the mountains there were less prosperous men who consistently voted against the agrarian interests of the Inland Tidewater. To those men, who were landowners and therefore voters, Cassius Clay would make an increasingly vigorous appeal. His efforts to win political favor by representing southern industrial interests were, he felt, more widely appreciated in the mountains. There the plantation-type, slave-worked agriculture did not predominate, and there minerals and water power made manufacturing feasible.

Clay recognized the difference of interest between the two sections and sought to win the support of the mountain men. "Now I . . . propose to educate a class to make capitalists 'of the manufacturers of American Switzerland . . . resting on *nine* States," he later explained, after many years of political effort. The mountains of the South were, he said, the "*greatest mineral district in the world.*" By educating mountaineers, "we shall retain of course our greater share of the uncounted mineral wealth for all time, otherwise foreigners will own it, and we will be their slaves." Much of Clay's activity after 1845 was directed toward instructing and guiding potentially industrial hill folk.[10]

Along with the political and economic matter in the reestablished newspaper appeared the recurrent strain of Clay's humor. In his wit, Clay made use of the new moral argument which he had adopted. "There are many men professing the Christian religion, who also profess to believe Slavery a Divine institution," he began, in his bitter sarcasm. But, he said, he had never heard a prayer offered for the holy bonds of slavery. "*If it is of God,*" he suggested slyly, "*Christians pray for it!*" He suggested a litany for the slave religion. "Oh thou omnipotent and benevolent God, who has made all men of one flesh, thou father of all nations," he prayed, "we do most devoutly beseech thee to defend and strengthen thy institution, American Slavery!"[11]

But no amount of earnest argument or biting satire could aid his cause in Kentucky. After August 18, he gradually lost the following he had built up. Clay's reputation grew in the North, however, where he appeared as a martyr to the cause of freedom. He did not wait long to profit by the publicity. In January, 1846, when he journeyed northward to deliver a series of lectures, he discovered that his notoriety had swelled his audiences. In Philadelphia's Musical Fund Hall, before the Board of Home Missions of the Methodist Church, he spoke on the topic, "Labor, the basis of the rights of property, cannot be the subject of property." His audience suggested the new departure into the realm of religious exhortation, but his subject indicated that he repeated time-worn arguments. Later in the same month he addressed an audience described as "the largest and most respectable concourse ever assembled under one roof in the city of New York," at Broadway Tabernacle. There he delivered an antislavery message to the "sympathizing thousands" who flocked to see the living martyr. In the opinion of contemporary critics, the speech he made there was his finest. Although it was a repetition of his economic and political ideas, he had polished it by long familiarity until it glittered with memorable phrases.[12]

In the climax of the Tabernacle address, Cassius achieved his most successful oratorical effect. As he concluded his tirade, he had completely awed his audience. He had to shout to force his voice to the edges of the crowd, but that extra strain did not diminish its trenchant, thrilling tone. The peroration fell upon the hushed multitude and inspired them with its patriotic fervor. After interminably expounding his antislavery arguments and painting himself as the suffering servant of the slave, he launched his colorful conclusion. "Come then, thou ETERNAL! . . ." he prayed, "inspire my heart—give me undying courage to pursue the promptings of my spirit; and whether I shall be called, in the shade of life, to look upon sweet, and kind, and lovely faces as now—or, shut in by sorrow and night, horrid visages shall gloom upon me in my dying hour—OH! MY COUNTRY! MAYEST THOU YET BE FREE!"[13]

Those melodramatic phrases so impressed his hearers, and his tales of suffering for enslaved humanity so affected them, that many who heard him remembered the magnetic effect after a half-century had passed. Nearly sixty years later, when a journalist penned Clay's obituary, he quoted those lines as the high point of Clay's oratorical career. Many of his listeners, in an era of forensic effusion, considered it the most effective speech they had heard.[14]

Clay could attract admiring audiences in the North, but he could not effect a political revolution in Kentucky. That, more than an avid northern following, was his ambition. In his own state, however, his efforts continued to bring fewer returns. Even in defeat he consoled himself that he had registered a victory. His newspaper continued publication, and a successor to it, the *Louisville Examiner*, was published inside the state. Four years later the Emancipation Party made a respectable showing in the constituent elections. Clay's clash with the proslavery party, though ending in defeat for his own plans, was not without its recompense.

But its effects lingered. As a result of Clay's immoderate essay of August 12, the slave party took advantage of the public excitement to unify opposition to him. The state legislature passed a law which severely restricted the discussion of slavery. Any person found guilty of "delivering to or disseminating directly or indirectly amongst the slaves any newspaper" or other document which might be construed as an attempt to produce insubordination, would be guilty of high misdemeanor. Moreover, any person who would "bring into contempt the lawful authority of the owners of slaves" would become subject to a fine of not less than five hundred dollars. The proslavery party would define just when a paper was insurrectionary, inasmuch as any newspaper published in Kentucky could be considered as "indirectly" disseminating its paragraphs among the slaves. The new law, said the editor of the *Observer and Reporter*, was entirely the result of Cassius Clay's "rash movement" in establishing among a dense slave population an outspoken antislavery journal. Clay's task, which before the suppression had not been a simple one, now became even more difficult.[15]

To reinstate himself in the eyes of those to whom he appealed, Cassius Clay needed to perform some noteworthy deed. In the Mexican War he discovered a golden opportunity to win friends in Kentucky. He did not hesitate to make use of it.

CHAPTER IX

TO THE HALLS OF MONTEZUMA

IT WAS an event far from Cassius Clay's editorial desk that offered the Kentuckian an opportunity to regain the confidence of his neighbors. Just across the muddy Rio Grande from the Mexican town of Matamoras, a detachment of United States troops under the command of General Zachary Taylor faced defending Mexican forces. The Yankees stood on disputed ground, and the Latins considered their presence an invasion. Suddenly, on April 25, 1846, after weeks of defiant indecision, Mexican cavalrymen sloshed across the river and engaged a party of United States dragoons in a skirmish which resulted in Yankee casualties. General Taylor immediately informed his commander-in-chief of the engagement. President James K. Polk, who had already decided upon war with Mexico over other matters, and who was at work on a war message when Taylor's information arrived, now had a provocative incident to report to the Congress. "After reiterated menaces," Polk told the lawmakers, "Mexico has passed the boundary of the United States, has invaded our territory and shed American blood upon the American soil." Polk demanded war with Mexico. Congress complied and approved a volunteer army and a war fund.[1]

Although many Americans agreed with Polk's indictment of the Mexicans, Cassius Clay joined the group which criticized the President's war message. As an extreme Whig and an antislavery spokesman, he opposed the war as an instrument of slave expansion, and as an American aggression. He charged that the war was the work of thieves. The motive of the war, he said, was plain enough: "It is plunder." The perpetrators of the war were eager to cross the Rio Grande to "glut their avarice, and flush the spirit of rapine." Clay regarded Polk's war message with a skepticism born of political convictions. "We doubt whether there has been any invasion of the territory we claim," he said, "and we feel confident that hostilities have not been commenced by the Mexicans." Repudiating the administration's version of the beginning of the war, Cassius placed the entire blame upon Polk and the Democrats. "We have not the least shadow of title to the land west of the Nueces," he declared. "We solemnly protest against the damning usurpation of James K. Polk in *making* war without the consent of Congress. . . . We demand of Congress, as a citizen of a republic . . . to cause the President to withdraw his forces from the soil of a friendly sister republic, and punish him for . . . making war without constitutional rights!"[2]

Although he opposed the war, in mid-June Clay offered himself as a volunteer soldier. His contemporaries expressed amazement at what they called an inconsistency. Clay, "with his foot in stirrup and his harness girded about him," as one of them described him, "pauses a moment . . . to heap maledictions upon the originators of that war, beneath whose standards he has volunteered to fight." But Cassius had a ready explanation: he claimed that despite his political views, he owed a citizen's allegiance to his government. "Resistance to it would be rebellion," he explained; "if general, anarchy . . . would be the result." The war, "so unjustly and wickedly begun," should be pursued with vigor. But his participation in the war did not signify a surrender of his political views, he promised. "Not a hair's breadth of sentiment, of opinion, or of opposition, shall we yield. . . ." Although he volunteered for the fighting, he did not relinquish his place among the opposition.[3]

Cassius proclaimed that his enlistment violated no tenet of his faith, but Yankee abolitionists continued to scoff at his action. The Ohio American Anti-Slavery Society, in its annual convention, denounced Clay for perpetuating slavery by taking part in the war. A religious antislavery journal mourned that he had departed the "true battlefield for the false one." The *Anti-Slavery Bugle* explained Cassius' fall from grace as an incomplete conversion. "He never *professed* to be baptized into oneness of feeling with the slave," the editor charged. Clay merely urged gradual liberation, "and that primarily on the ground that it would advance the interests of Kentucky, and benefit the *Anglo-Saxon* race." While Clay's influential neighbors continually associated him with the abolitionists, the Ohio antislavery editor rejected the inference.[4]

Though Yankee abolitionists belabored Cassius for entering the war which he had criticized, their censure did not deter him. In the war he had seen a chance to reinstate himself in the eyes of his neighbors and to overcome the ill effects of his editorial error of August 12, 1845. His participation would prove, he said, that an "honest avowal of an eternal war against slavery, did not of necessity deprive one of the confidence of the *people* of our noble State, however much the slaveholders might denounce him." In going to war, he said, "I wished to *prove* to the *people* of the South that I warred not upon *them* but upon Slavery, that a man might hate slavery and denounce tyrants without being an enemy of his country." "He believed," an apologist explained, "that [by enlisting] he would be enabled, on his return, to discuss the question of emancipation freely. . . ." To counteract opposition to his political program, Cassius had decided to volunteer.[5]

Clay packed his military equipment and prepared to march to the Halls of Montezuma to prove his patriotism, and for other reasons, too. He possessed an adventuresome spirit, and he yearned to escape what he called the "dray-horse duties of Editor." Moreover, he had an established career as a citizen-soldier, with a commission as colonel of state troops, and he loved the life of the militiaman. "It was his boyhood ambition to figure as a soldier," one of his close friends remarked. "He looked upon the tented field, and its pomp and panoply, as a goodly and a glorious thing." Leaving his business affairs with his brother Brutus, and the newspaper with his associate, John C. Vaughan, Cassius eagerly prepared for war.[6]

As a colonel in the state militia and commander of its annual encampments, Cassius expected that he would receive an invitation to command one of the Kentucky volunteer regiments. He soon learned, however, that he would receive no commission for the war, and he believed that he had been rejected because of his political views. One of the influential slave-owners, he complained, had told the governor that if Clay were "*elected to any command, and goes to Mexico, he will triumph over us, in spite of all we can do*." The complaisant governor then refused to make use of Clay's talents. Undaunted by the snub, Cassius announced that he would "fall into the ranks, as a private, with my blanket and canteen." [7]

He enlisted in the "Old Infantry" company, a Lexington militia unit older than the state of Kentucky. Years before, as he had advanced in the militia, he had been its commanding officer. Now the company, under the command of Captain James S. Jackson, voted to volunteer for one year as a unit and decided to mount itself to join the Kentucky Volunteer Cavalry Regiment. Though Clay "fell in" as a private soldier, he had his own plans for winning military promotion. He brought with him several thousand dollars and let it be known that if he were its leader he would handsomely outfit the entire company. Cassius induced Captain Jackson to resign his office and to throw the captaincy open to election by the soldiers. They unanimously chose Clay as their leader. Cassius attributed the honor to his popularity and to his military ability—he called it later the "greatest honor ever given an American citizen." But his money also had some obvious effects. With their new, colorful regalia, Captain Clay reported that his men belonged to the "brag com.[pany] in camp." Again he boasted that he had foiled his enemies; their attempt to discredit him had failed.[8]

After a week of drill on Clay's lawn, the "Old Infantry Cavalry," as they incongruously styled themselves, prepared to leave their home station. At noon on Thursday, June 4, Cassius formed his men, mounted on fine Kentucky horses, with two other cavalry troops from the vicinity. Patriotic Lexington citizens planned a public ceremony at their leave-taking. The ladies of the Female Bible Society contributed to the spiritual welfare of the expedition by presenting a copy of the Holy Book to each departing trooper. A shower of rain mercifully spared the volunteers a lengthy sermon from a local minister, who contented himself with a prayer.

The ceremonies over, orders rang out, and the mounted men moved off through the rain, headed for Louisville, where they would join the regiment. Captain Clay proudly rode at the head of his column on the way to new adventures.[9]

If Cassius considered the dangers which lay ahead, his dark eyes gave no evidence of it. He left the worrying for Mary Jane, who had ample cause for concern. She took responsibility for their three children, and while her husband was in Mexico she would bear him another son. In addition to her family cares, she assisted in the management of the White Hall estate and Cassius' other enterprises. And as if that were not burden enough, she had reason to worry about her marriage. Although for years there would be no open break between Cassius and Mary Jane, an estrangement was building up. Years later, when Cassius described the steps which led to his broken home, he recalled that in 1845 he had thrown away his wedding ring following a quarrel. Nevertheless Mary Jane loyally accompanied Cassius to Louisville, where she stayed until the regiment left the state.[10]

In Louisville the men of the regiment lived in tents on the bank of the Ohio River. There, on June 9, they were mustered into government service and received a clothing allowance, partial pay, and their arms and ammunition. During the days of preparation they encountered the mismanagement with which the army met its war crisis. Friends from Lexington who witnessed the muster returned home to criticize the "unpardonable neglect" which had resulted in a complete lack of the "absolute necessities" for men and horses. Many times in the coming year the men would echo—in more colorful language—that criticism.[11]

For the entire month of June the men of the cavalry regiment remained in Louisville, awaiting orders and transportation. Public ceremonies improved their morale. They paraded through the streets and received the praise of the uncritical spectators. Captain Clay added to the unit pride by presenting his company flag as the regimental colors. The "Old Infantry" flag represented gallant action in the War of 1812, he pointed out, and was a fitting emblem for the regiment. Colonel Humphrey Marshall accepted the flag with the promise that the honored talisman would be well protected. Marshall, a West Point graduate who had resigned from the regular army, had, in 1836, organized a volunteer company which marched to the Texas frontier to aid in the revolution against Mexico. That expedition had prepared him for the task ahead. As the commander of a volunteer unit he would have difficulty with fellow officers who attempted to use his regiment to further their own ambitions. Long before the year of enlistment had expired, Marshall, like the men he commanded, would have had enough of the war and its petty politics.[12]

Although in a few months the soldiers would long for their home state, in the beginning they eagerly anticipated battle. But there were some among them who regretted their enlistment from the first. On the night before the regiment embarked for the battle zone, several reluctant troopers deserted and took refuge in a house of ill fame on the lower end of the river town. Colonel Marshall ordered Captain Clay to take a squad of men and arrest the deserters. When the Madame refused to unlock the door, Cassius ordered his men to force an entrance. To resist them, the deserters fired several volleys from the windows, hitting one man in the face. The shots did not stop Clay's men, however, and they broke into the house. In the melee which followed, the house and its furnishings were badly damaged, but Captain Clay triumphantly returned to camp with the deserters.[13]

Finally, with all the volunteers in the fold, the long wait ended. Early in July, the troopers struck their tents, rolled their blankets, and then, after a forty-eight-hour wait at the landing, boarded steamboats bound for Memphis. Cassius had bidden Mary Jane farewell a few days earlier, and she had returned to Lexington. Along with the thousand men of the regiment, he began the long trip to the combat zone. It was a journey which the Kentuckians would long remember, for they endured countless hardships in the three months which followed their departure from Louisville. On the river, a violent storm endangered the boats and frightened the men. And every day, despite all precautions, a few soldiers managed to fall overboard. On July 7, the men arrived in Memphis, where they unloaded and set out on the long march overland to San Antonio. In the hot months of the year they traveled through the humid Arkansas swamps and traversed the stifling bake-oven which was the Texas plain.[14]

Though the Kentucky soldiers blamed the army for their suffering, their own weakness added to their pains. Unaccustomed to the rough life of the trooper, they were unprepared for the physical requirements of war. But the army made no effort to ease their path. The men complained that the route they took was the worst possible way to south Texas, and they were particularly aggrieved since in the midst of their trek they were rerouted to Port Lavacca on the Gulf, where they could easily have been sent by water. It was a "remarkably fatiguing trip of 8 or 900 miles, through a burning sun at an inclement season of the year," one soldier reported. On some days they went without water altogether; at other times they rejoiced to drink what they scooped from a brackish hole.

The volunteers also complained of the quartermaster's "negligence or oversight," which necessitated long, hungry marches for supplies. They had ordered rations and forage to be delivered to them along the march, but their requisitions were lost in the confusion of army administration. And they grumbled about their clothing. The resplendent outfits which had brightened the Louisville parade proved unsuitable for the wilderness. One soldier lamented that the men had been "*absolutely turned naked in a wild country*." When their complaints became a political issue back home, the army answered that the regiment had received its clothing allowance in Louisville, and if the men were without clothing within two months they had only themselves to blame. A Lexington private admitted that the reason the men had no uniforms was that they had "traded all . . . superfluous clothing long since for whiskey, potatoes, and the other necessaries of life. . . . My unmentionables," the fellow joked, "give evident symptoms of an intention to desert me in my *extremity*."[15]

The Kentuckians also blamed the army for the alarming number of their men who fell ill. At every stop along the route they left stragglers who could not keep up the pace of fifteen miles a day. After they had arrived at Port Lavacca, the surgeon listed 160 new patients on one day. "*We are sick*," a soldier diagnosed, "*of an order from a General who don't know what he is about*." Long before the regiment saw any action, it had lost one-half of its effective strength from the punishing overland march. Its personnel confessed that with the alleged blunders and the needless suffering, they had endured enough of army life. When their year of enlistment expired, although the war had not been settled, the Kentuckians packed their duffel and returned home.[16]

The road to war was filled with tribulations for the recruits, but Captain Cassius Clay enjoyed the excursion. When the regiment crossed into Texas he received permission to leave its line of march and go on a buffalo hunt. Taking a friend from another company, he went into the Comanche country west of Austin. For nearly a month the two men roamed the plains, away from the outfit. In that wild country, the hunt was a struggle for survival. For several days they found no water, and the food was not always plentiful. They had a close call with some unfriendly Indians but escaped because of the speed of their fine mounts. Again, they lost their way and were given up for lost by the regimental officials, but at last they located their unit. For years afterwards Clay loved to recount, with added gusto, the tales of his buffalo hunt. In 1885, when he wrote his memoirs, he devoted nearly an entire chapter to a description of the exciting side trip.[17]

Upon his return to the regiment he participated in another off-the-record incident. Thomas F. Marshall, the lawyer who had delivered the oration at the suppression of *The True American* the year before, was also a company commander in the Kentucky cavalry regiment, and he and Cassius had quarreled along the way. One night, as they sat around the officers' campfire, Cassius entertained the group with a poem from his school days. But his memory was faulty, and he forgot one line. It ran, "When Greece her knees in suppliance bent . . . ," but he could not remember it, and repeated, "When Greece her knees, when Greece her knees. . . ." Captain Tom Marshall, with a humorous twinkle, asked him, "What do you want to grease her knees for, Captain?" The other officers set up a whoop at the ribald pun. Cassius, who had not forgotten Marshall's part in the attack upon his journalistic enterprise, considered his pride wounded and challenged the quipster to a duel. The next morning, as the regiment took up its march, the two men rode off with their seconds. In a secluded spot they fired but missed. Although Cassius constantly practiced his marksmanship, he had a poor record as a duellist. Fortunately for him, so did his opponents.[18]

When the cavalry regiment reached the war zone, Cassius continued his search for adventure. Regarding the war as an opportunity to build up a reputation of which he could make political capital, he volunteered for dangerous missions. He dashed off to the headquarters of several generals, begging an assignment which would provide glory as well as excitement. In the day when the efficient officer was the one who provided his men with action, Cassius became a popular captain. He hounded General John E. Wool for a job, and Wool, who had commanded the long march from Memphis, agreed to detach Clay's troop for an expedition to Chihuahua. When Cassius returned to camp with the prized order, however, he had to turn it down because his men were still too weak from their long journey.[19]

Clay rejected the special mission for another reason, too. He discovered that his efforts to see action had angered his regimental commander, Colonel Humphrey Marshall. When Marshall, who jealously guarded his unit, received the order detaching Clay's troop, he raged at the attempt to dismantle his command. He charged another volunteer colonel, Archibald Yell of Arkansas, with attempting to use elements of the Kentucky regiment to promote his own ambitions. To jealous American officers concerned with their own advancement, the least important enemy in the war was the Mexican Army.[20]

Although his politicking produced internal conflict within the officer corps, Cassius was finding that it had the effect he desired. Gradually he won the confidence of his comrades. "I find that this gentleman, who had gained an unenviable notoriety by his mad and *selfish* course on the slavery question," one of the Kentucky soldiers reported, "is acquiring, by his strict discharge of duty, more standing as an officer than probably any other individual in the regiment." Clay's effort to use the war as proof of his patriotism was bringing results.[21]

He missed no opportunity for action and pushed himself into as many assignments as he could. In mid-December a Mexican traitor reported an armed band a few miles away. When orders were issued for the Kentucky cavalry to dispatch an investigating patrol, Captain Clay implored his colonel to give him the mission. Before the detail could leave, however, a messenger brought news of a hot fire-fight between the regiment's supply train and enemy marauders some fifty miles away. Additional men strengthened the original detail, and they galloped to the rescue, with the yelling Cassius in the lead. One of the troopers who made that wild ride declared that he would never forget it. They left the camp at dark, he recalled, and by daylight had covered the entire distance—generously estimated at fifty-eight miles—"through the thickest chaparral." When they reached the train, saddle-weary and thicket-whipped, they were disappointed to learn that their speedy dash had been in vain. The fight had ceased hours before. But by such action, and by his exuberant love of the trooper's life, Cassius became the regiment's best-liked officer. The men recognized that if they wanted to enjoy hard riding and exciting missions, they would have to follow Captain Clay. With such a reputation, Clay took part in an expedition which proved disastrous.[22]

On January 19, 1847, Cassius and thirty picked men left camp with Major John P. Gaines in command. Their announced purpose was to hunt fodder, but in reality it was to probe enemy strength on the Saltillo front, commanded by General William J. Worth. For three days they rode southward over the plains beyond Saltillo but found no enemy. On the afternoon of the third day, they arrived at the hacienda of Encarnacion, about forty-five miles south of Buena Vista. There they met another United States scouting patrol, commanded by Major Solon Borland, a medical doctor from Arkansas. The Arkansawyers told the newcomers of a band of enemy soldiers reported to be in Salado, some distance further south. Together the two parties rode toward that city, but darkness and the approach of a storm forced them back to the hacienda for shelter.

Encarnacion, similar to many another Mexican hacienda, was a large stucco building with a flat roof, surrounded by a wall. The cavalrymen corralled their horses in the courtyard and settled down for the night. Ever afterwards, Cassius insisted that he had protested the bivouac in the hacienda, but that he had been outranked. He also contended that the two majors had vetoed his suggestion that a picket-guard watch the main roads leading to the farmhouse. They posted only night watches on the roof of the house, and with that inadequate precaution, went to sleep. During the night one of the sentinels gave the alarm, but the sleepy men explained the noise as someone drawing water from the courtyard well.[23]

When the morning mists lifted, however, the men discovered that their fortress was entirely surrounded by Mexican cavalrymen. The trap had sprung upon seventy-two Arkansas and Kentucky troopers. Despite the obvious disparity of the forces, they prepared to resist, determining to make Encarnacion a second Alamo. Barricading themselves behind the walls of the house and planning to make every shot count, they watched as the enemy closed in upon them. But as they aimed their opening shots, they saw a white flag approaching with a deputation to receive their surrender. After several hours of indecision, the outnumbered force decided that resistance would be futile. At 11 A.M., January 23, 1847, they surrendered themselves as prisoners of war. Although they estimated that they were outnumbered forty to one, many of the stalwart men wept as they stacked their unused weapons in the hacienda courtyard and walked out to surrender.[24]

Captain Cassius Clay, who had sought glory in the war to further his political career, became a captive. Before any major action had developed, the war had ended for him. With his penchant for enthusiastic volunteering, he missed the big action and became one of the few American officers who fell into enemy hands. A month later the men of his regiment played a decisive role in the victory at Buena Vista, but he was on the way into Mexico under guard.

Despite Clay's apparent failure to win renown as a soldier, he received widespread acclaim for his conduct as a prisoner. The day after their surrender, one of the captives escaped, and in the excitement Cassius' quick action saved the lives of the other prisoners. Among the Arkansas troops there was a guide named Dan Drake Henrie, who had been captured by Mexican officials in a prewar foray across the Rio Grande. He had escaped then, but at the Encarnacion surrender he had been recognized, and he feared that he would be shot. Under the guise of checking the line of marchers, Henrie dashed away before the guards could draw their carbines. Though in the Encarnacion fiasco seventy men were lost to the American fighting force, the message Henrie delivered to the American commanders at Saltillo was sufficient recompense. When Henrie informed General Taylor of the advancing Mexican Army, the general took personal command at Saltillo and prepared the battle orders for what developed into the Buena Vista engagement.[25]

Henrie's escape was a boon to the Yankee cause, but it resulted in a tense moment for the remaining captives. The surprised guards supposed that all the prisoners would bolt when Henrie did, and the Mexican commander ordered his men to lance the captives. As the Mexicans, levelling their sharp spears, charged upon the defenseless troopers, Cassius Clay's long experience with danger saved them. Quickly he ordered the prisoners to lie down and to make no show of resistance. Then heroically—and with typical melodrama—he bared his breast to the onrushing lancers. "Don't kill the men: they are innocent," he shouted. "I only am responsible." His orders to the men, who instantly obeyed, and his own unusual action stopped the impending execution and allowed him time to explain. In broken Spanish and in English, Clay said that Henrie had escaped for his own reasons and that the men had known nothing of his plans. After some minutes of debate, the officers of the guard were convinced that no mass break was intended, and Cassius as well as the others was spared. Although his enemies circulated derogatory versions of the incident, Clay's act brought him much favorable comment. "Who but C. M. Clay, with a loaded pistol to his heart, and in the hands of an enraged enemy, would have shown such magnanimous self-devotion?" his fellow prisoners asked. "If any man ever was entitled to be called 'the soldier's friend,' he is."[26]

After surviving that near-disaster, the prisoners resumed their long march southward, the men on foot and the officers mounted. Cassius shared his mount with the footsore men, walking so they might rest occasionally. He also supplied money to those who needed it and "resorted to every sacrifice" to make the men comfortable. The group remained for some weeks in San Luis Potosi before going on south.

Throughout the journey they suffered from lack of water and received little food. They were convinced that they ate dog or mule meat, since they saw no commissary stores anywhere. As they passed the widely separated ranches, however, the women ("in all countries the most charitable," Cassius gallantly declared) would run out and offer them eggs, dishes of beans, and the native tortillas. But in the towns the irate populace would gather to mob them, and they often encountered showers of stones. Their guards were kept busy preventing even worse treatment.[27]

When the weary prisoners finally arrived at the City of the Montezumas, they found an insurrection in progress. They were kept outside the city until the end of the day so that the guards would not themselves become prisoners of the wrong side. After dark they were smuggled into the city and taken to a state penitentiary for safekeeping. The prisoners considered it an insult to be incarcerated among the "common felons," but they remained there for the months they stayed in the city. They complained also that they should have been immediately exchanged. Many Mexican soldiers and officers had been taken at Buena Vista, so that there was sufficient basis for an exchange agreement. No Mexican official would take the responsibility for negotiating a protocol, however, and the men remained in the capital of the enemy country. After they were given their paroles, they spent much time sight-seeing in the beautiful city.

Despite that courtesy, the men still fumed about their prison conditions. They had no beds and slept on the floor with only their horse blankets for warmth. They received no clothing—the Mexican quartermasters emulating their United States counterparts—and food in such small quantities that they had to purchase additional rations. The Mexican government paid them only fifty cents a day, but they borrowed from a New Englander who had settled in the city. "Living in this city is higher than in any place in the world," Cassius reported, "and in consequence we are somewhat troubled to get the means of support." The embattled Mexican government, split by an internal revolution and invaded by the United States forces, did the best it could to provide for the prisoners, but the Yankees did not consider their arrangements satisfactory.[28]

While the men protested their treatment, Cassius continued to denounce the war effort. As he had promised at the beginning, his participation in the war would not blind him to its blunders or to its injustice. From his prison cell he fulminated against the politicians who managed the war. "Can any man tell me why all this expenditure of blood and money?" he demanded. "Have we not land enough? Do we want eight millions of revolutionary Indians and half breeds to increase the difficulties of the elective franchise . . . ?" He blamed the Democrats for the tragedy of the war and urged that the Whigs oust them at the next election. "I hope they will everywhere be defeated," he said. "They have carried their party feelings into the military appointments and attempted to disgrace eminent men who were risking all for their country . . . is not such conduct sufficient to disgust all men?" Cassius looked to the day when he would again take the stump against his political foes.[29]

Penning political diatribes against his opponents, sightseeing, and recovering from an attack of food-poisoning, Cassius spent his days in the Mexican prison. Although there were constant rumors that they would be exchanged, the men remained in Mexico City for eight months. As the main American offensive under General Winfield Scott approached the city from the east, however, the prisoners were removed to Toluca, a pleasant provincial town and the capital of the state of Mexico. During the move to Toluca, some of the officers, including Major Borland, escaped to the United States lines. Cassius righteously refused to accompany them, proclaiming himself still bound by his parole and concerned for the welfare of his men. At Toluca, the prisoners lived in a monastery which served them as an asylum from the vindictive citizens. Guards were placed around the building, not to keep the men in, for they were on parole in the city, but to keep the Tolucans out.[30]

In spite of the dangers of moving about among the aroused citizens, Cassius determined to "enter society as far as I was able." He was especially attracted to a young Tolucan senorita, eighteen years old, named Lolu, and he braved all obstacles to visit her and her pet parakeet, Leta. He knew that if he were identified as an enemy soldier, he would be instantly killed by the Tolucans. Attiring himself in a colorful serape and a wide-brimmed sombrero, Cassius disguised himself as a Mexican and thus passed through the crowded streets. Lolu, he boasted years later, enjoyed his company and his conversation. When he was much older he described his acquaintance with Lolu as one of the "real joys fading into the dead past," which left "rose-tinted memories of the days that are gone. . . ." He had been promoted, Cassius claimed, from prisoner to conqueror.[31]

At length the interlude, which Cassius considered a luxurious respite after his "exhausting use of the nervous powers" of the prewar years, came to an end. After months of negotiations, the prisoners were released in late September and sent to Tampico for exchange. Two Mexican generals were offered for the three American captains. When he arrived among the United States troops, Clay learned that the men of his regiment had arrived in New Orleans in early June and had been home for some months. He learned also that the unit had participated in the battle of Buena Vista and had suffered casualties of nearly twenty per cent while taking a major role in the engagement.[32]

Cassius was delayed in leaving Mexico, so that he did not rejoin his family until December 6. Mary Jane, with their four children, met him at the home of his mother in Frankfort. Back in Kentucky, Cassius discovered that his prison experiences had alleviated much of the angry opposition to him. On October 25, before he returned, the Whigs of Estill County met and endorsed Major John P. Gaines for governor and named Captain Cassius M. Clay as their choice for lieutenant-governor. Previously, Major Gaines had been elected to Congress from Kentucky's Tenth District, but the effort to elevate Cassius made no further progress. Clay received numerous complimentary dinners and was invited to deliver addresses about Mexico and the war. For a brief period many of those who had participated in the attack of the eighteenth of August joined in eulogizing him. With the commendatory resolutions ringing in his ears, and with an ornate gift sword gracing his mantelpiece, Cassius prepared to profit by the good will which his war record provided. With the national election of 1848 and the state constituent election of the following year coming up, there was the chance that, as a heroic soldier home from the wars, Cassius Clay might succeed in his ambitious undertakings.[33]

CHAPTER X

CLAY ATTACKS THE WHIGS

CASSIUS M. Clay was as displeased with Kentucky political parties after his return from Mexico as he had been earlier. Now thirty-eight, with his handsome face beginning to show the engravings of an unquiet life, he still pursued the dream of political power. But the old parties offered him no better opportunity in 1848 than they had two years before, and he continued his efforts to create a new organization in which he would gain his ends. He deplored the increasing proslavery pressure which drew Whigs and Democrats together, and he declared that there was a need for a new opposition party. To dramatize the similarity of the two established groups, he planned to defeat the Whigs and thus reduce the proponents of slavery to one party. "It was a part of my policy to destroy the old parties," he recalled later, "to build up a new one of universal liberty." To accomplish his purpose, Cassius Clay waged war upon the party of his youth.[1]

His first step was to discredit the state's Whig leaders. Chief among them was the Great Compromiser, Henry Clay, who for years had reigned supreme in Kentucky politics. Even before the war Cassius had challenged Henry's lead, and for three years he had nursed a grudge against his kinsman. In the spring of 1848, for political purposes, he resurrected his anger. He declared that on August 14, 1845, just before the suppression of *The True American,* Henry had hurriedly departed Lexington, "leaving your friends and family to murder me, for vindicating those principles which you had taught me, in your *speeches at least.*" For that treachery, he asserted, Henry no longer merited high office. "I would never silently see a man elevated to the Presidency of the States," Cassius announced, "who winked at the overthrow of the . . . press." But he was not content with a denial of Henry's political suitability. With characteristic spite he proclaimed himself the personal foe of the Gallant Harry. "I ceased to be your friend," Cassius told him, "and became, by the necessity of my nature, your enemy. . . ." His belligerent personality allowed him no middle ground between friendship and a hateful enmity. His remarks, one commentator said, were so "bitter and ill-tempered, as to be repulsive to . . . right-minded men of all parties." [2]

Cassius was not alone in his repudiation of Senator Clay. Many Whigs had come to the conclusion that their aging leader—despite his persuasive charm—had too many enemies to win a national election. "I am tired of being beaten," a Whig explained. "I am in favor of Mr. Clay, . . . [but] if we persist in his nomination, defeat is certain." Even state Whig leader John J. Crittenden agreed that Henry would be a liability to the party. "I prefer Mr. Clay to all men," he was careful to say; but added, "My . . . involuntary conviction . . . is that he cannot be elected." Desperate for the spoils of political victory, men like Crittenden were willing to turn to a glittering military hero, General Zachary Taylor, who, though a newcomer to politics, was triply recommended as a slave-owner, a professional soldier, and a planter. Carefully scrutinizing the political straws, Cassius Clay discerned the strong inclination among Kentuckians to jettison the master of Ashland in the hope that Taylor might unify their divergent interests. With all this in mind, Clay staked his claim to a seat among Kentucky's rebellious Taylor Whigs. He intended to do his share to win Kentucky support for Taylor, not only to back a winner, but also to embarrass the state's Whig leadership. For if Henry Clay failed in his own state, he stood little chance at the national party convention.

The Whig meeting at Frankfort was, therefore, significant. The party members gathered on Washington's Birthday, but so deep was the rift between the supporters of Henry Clay and those of Taylor that they met in two separate groups. The Taylor forces were more numerous, and Cassius joined those men like his old regimental commander, Humphrey Marshall, who demanded that the state's entire vote be given to the general. Taylor's strength was so great that the Clay Whigs would not permit a vote for fear their candidate might lose. Finally a compromise healed the breach: one of the senatorial delegates to the national party convention would be committed to Henry Clay, the other to Taylor.[3]

Having successfully forced the Clay Whigs to recognize Taylor's strength, the dissident wing of the party met and pledged itself to secure the general's nomination. Cassius Clay publicized the action of his colleagues—thereby making himself appear to be the ringleader of the Taylor movement. When the state convention adjourned, Cassius wrote an open letter to a New York pro-Taylor paper, declaring that Henry Clay's own state party had repudiated him. In the name of the entire Kentucky Whig Party, Cassius brashly announced that the state "cherishes the long service of Henry Clay, but also believes that Mr. Clay cannot be elected." He added that further attempts to nominate Clay would be futile. With that interpretation Henry Clay's friends violently disagreed. "The Clay-Whig press roared as a herd of wild beasts," Cassius chortled many years later, thoroughly pleased at the memory of the commotion he had caused. But the outcry was to no avail: at the Baltimore convention General Taylor won the nomination, and as the years passed, Cassius Clay came to consider that he was entitled to the credit for Taylor's success.[4]

Cassius had chosen the winning contender in the race, but he took little part in the election because of his personal quarrel with the Whig Party, whose members had joined the Democrats to attack his press. In addition, he was becoming more sympathetic with the Liberty platform. "I am pleased at the result of the Van Buren movement," he confided to his friend Salmon P. Chase. "I hope by another presidential year that *we* will come together." The Taylor candidacy, he said, was effecting that result: "Taylor will be elected, which will drive more of the democracy into antislavery formation, also many Whigs." Clay thanked Chase for suggesting him as a possible candidate in the northern movements, but for the present he would "stand for old Zack." And at home, he expressed confidence in the future. "The cause in Kentucky is steadily advancing," he said, "and my hopes are high for the life-time war."[5]

At this juncture the Kentucky General Assembly provided Cassius with the basis for a renewed battle for constitutional emancipation. When the legislature repealed the Negro Law of 1833 and opened the way for an unlimited increase in the slave population, many Kentuckians became alarmed. "Many I think regard the crisis as at the door," Kentucky Congressman Richard French told his Georgia friend Howell Cobb. The Reverend Robert J. Breckinridge, a Lexington minister long connected with the emancipation movement, declared that the legislative action had frightened "*perhaps* the bulk of the state," who favored no increase in the slave population. Cassius Clay rejoiced at the widespread criticism of the repeal. "The last legislature put its leaden heel upon us while we slept," he asserted. "Thank God! the touch of that heel has broken our slumber." So concerted was the protest that the assembly reluctantly voted to call a constitutional convention for October, 1849, to settle the question of slavery in Kentucky. That summer, when voters chose delegates, and when discussion of emancipation was widely acceptable, critics of slavery had their most auspicious opportunity.[6]

At the same time, however, the situation threatened to break out in violence, for any discussion of slavery and emancipation in the South was fraught with danger. Cassius had long worried about his friend Breckinridge, who had no basic training in personal combat and who lived in constant danger—or so Cassius thought. One evening he went to see Breckinridge to warn him of the hazardous times ahead. The reverend gentleman, Cassius announced, should be prepared to meet an attack should one develop. He had designed a weapon, he continued, especially to meet the needs of a man unused to personal combat, and he had ordered it manufactured in Cincinnati. Then he drew from his carpetbag the weapon—a knife with a blade seven inches long, and two inches wide at the hilt. This knife was unusual in more than mere appearance. It was intended to be worn strapped under the left arm, hanging upside down, with the knife held in its scabbard by a spring device. When the good doctor grabbed the handle, the spring released the knife at "belly level"—an attack no assailant would expect. Then, Cassius demonstrated, all Breckinridge had to do was "point the instrument at his [opponent's] navel and *thrust* vigorously!"

Years later, when the gentle Breckinridge showed the ferocious blade to his youngest son, he confessed that he always felt a little uneasy when he wore it. "Every time I gestured heavenward," he said, "that infernal knife thumped against my ribs!"[7]

Cassius planned a more thorough battle than one fought with the cold steel of vicious knives, however. Before the legislature issued its call for a constitutional convention, he had moved to organize Kentucky emancipationists. He had called a meeting of the "friends of emancipation" to be held in Frankfort to protest the repeal of the Negro Law. When the General Assembly acted, therefore, his plans were already made. He merely enlarged his invitation to include all who wanted the new constitution to provide for gradual emancipation. Clay confidently expected the Frankfort meeting to produce the emancipation party of his dreams.

On Christmas Day, 1848, he wrote out his aspirations for the new party. He desired a fully developed organization, with a treasurer and committees of finance and correspondence, and he envisioned a districting of the state, with orators allotted to each county. The next step was to encourage county meetings to name delegates to the state emancipation convention. Such gatherings met in about one-fourth of the state's counties. On April 2, 1849, the "friends of gradual emancipation" in Clay's home county convened at the Methodist Church in Richmond, where they elected state delegates and planned an independent party.[8]

The repeal of the Negro Law of 1833 and the subsequent call for a constitutional convention brought many respected Kentuckians into the field as gradual emancipationists. In the Fayette County meeting, for example, Henry Clay presided, and Robert J. Breckinridge presented the resolutions. Breckinridge proposed a program as ambitious as that of Cassius Clay. "Whatever we go for, will be called emancipation, and resisted as such," he asserted, "while, if we go for all we desire, our opponents, in order to defeat us, will be apt to concede to the popular wishes . . . and may come up, even to our second ground." As time went by, it became evident that Breckinridge's analysis was sound.[9]

Both the impetuous Clay and the more moderate Breckinridge planned comprehensive programs to present to the state convention. They soon discovered, however, that most of the delegates were too moderate to accept their views. The convention of the Friends of Emancipation assembled in Frankfort at 11 A.M. on April 25, 1849, with the conservatives in the majority. From all parts of the state had come about fifty delegates, characterized by "much respectability and much talent." Most of them were Whigs, nearly three-fourths were slaveholders, and about one in seven was a minister. Cassius Clay, who had voluntarily abdicated from the slave-owning class, and who was not primarily motivated by humanitarian ideals, led the extremists. Though in the beginning he attempted to make friends by reconciling his views with those of the majority, it was not easy for him to do.[10]

As the leader of the humanitarian reformers, Robert J. Breckinridge delivered the introductory address. Involuntary hereditary slavery, he said, was detrimental to the prosperity of the commonwealth, inconsistent with free government, injurious to a pure state of morals, and contrary to natural law. But he recommended that the delegates approve no plan to end the evil without compensation, and he suggested that they confine themselves solely to unborn Negroes. When Breckinridge finished, Cassius Clay followed him upon the stand and spoke without unduly shocking his hearers.[11]

In his eagerness to have the emancipationists present an undivided front, he accepted the mild report of the resolutions committee that the new constitution should merely recognize that the citizens "had power to enforce and perfect a system of gradual prospective emancipation of slaves." Since Cassius had assumed that right for years, he said that the statement was too weak, but he agreed to support it to keep the peace. "My judgment tells me," he explained, "I must yield to the maturer and better judgment of the majority." In his effort to win friends, he temporarily sacrificed his ambitious program.[12]

He concurred in the committee report but he opposed an attempt to weaken the resolution still further. Judge Samuel S. Nicholas, one of the older slaveholding emancipationists, proposed alternative resolutions which denied all Clay's past efforts and would have tied his hands for years. Nicholas suggested that the new constitution allow elections "on some future day," to decide whether a scheme for gradual emancipation should become a part of the new code. Until that time, "to prevent the injurious effects of a constant, or even too frequent agitation of the question," the judge recommended that the emancipationists should not discuss the issue at all. Nicholas suggested that they not even advocate their views as a feature of the new constitution, for he feared that they might lose everything by gambling on the elections. The proslavery party—"I mean perpetualists"—he said, was all-powerful. "They are bonded with both parties and the wealth of the state." [13]

But Cassius would not agree to such a denial of his course. Even before Nicholas had finished, he was on his feet demanding the floor. When he received permission to speak, he ripped off his polite mask of congeniality. "I know," he began, "that . . . I am characterized as impulsive, hot-headed, reckless, and passionate." Yet he had, he said, made an effort to agree with the moderates. "We *fanatics* are willing to take your compromise," he told them. "We think it too moderate, and I have been reproached by some because I have yielded." But he would accept no further emasculation of the statement. "We who have, it is feared, compromised too much already, are asked to come yet lower down!"

Clay implored his colleagues not to prohibit discussion of the emancipation issue. "What if it be true that the politicians and the money power be against us?" he demanded. "Will our silence bring them to us?" He begged the emancipationists to join him in forming a new party because the perpetualists had "bonded" with the others. "Have not the old parties forgotten their allegiance to the right in all things," he asked them, "to fasten upon the country this curse of Slavery?" Already, he said, a new movement was underway. "The party in favor of freedom is growing everywhere. It has broken through party restraints at the North. It will do so here." He considered the issue too pressing to postpone. "For myself, I am for agitating this question," Cassius, after a decade of such activity, declared. "We must convince the people—the real people—of its importance before it can be done. . . . We must seek them out—at the crossroads and places of public resort in their neighborhoods. We want men on the stump. We want to get at the ear of the people." Under the influence of Clay's impassioned oratory, the delegates rejected Nicholas' proposals and opened the way for discussion of emancipation in the forthcoming constitutional elections.[14]

Cassius Clay took his own advice and carried the campaign directly to the voters. Packing his clothing and a brace of pistols in his carpetbag, and traveling in a light buggy, he spoke widely in favor of emancipation candidates. He continued to base his arguments upon the economic competition to free laborers and not upon the injustice of slavery to the Negro. "Ninety thousand of the 117,000 voters in the present election are dependent upon their own labor for subsistence," he would repeat. To those people he gave warning that slaves were beginning to compete with skilled laborers for jobs. If printing offices were "overrun with *Black Rats*, as are several of the mechanic shops in the interior towns," he declared, "we should not see the great body of the newspapers of the State in opposition to Emancipation." He reminded his hearers that in 1837, when the carpenters and painters of Louisville asked for a ten-hour day, slaves were employed in their place. "Ultimately," he concluded grimly, "slavery triumphed over freemen."[15]

With that message Cassius Clay spoke at numerous gatherings over the state, but his main interest was in the Madison County campaign. There the proslavery party nominated Squire Turner, William Chenault, and James Dejarnatt for seats in the constitutional convention, and the emancipationists chose Major Thompson Burnam. The candidates agreed to appear together and to apportion the time equally. Cassius, who often spoke for Burnam, disliked Squire Turner, the leading perpetualist candidate, and the two men had several quarrels. In mid-June their dispute erupted into violence, and bloodshed discolored the electoral process.[16]

On June 15, after several days of rest at White Hall, Clay started out on a week's speaking tour through the southern section of Madison County. He arrived at Foxtown, a village on the Lexington-Richmond turnpike about a mile from his home, just before another round of speeches was to begin. Turner took the platform first, and as he spoke Cassius interrupted him twice to "make an explanation." Some ominous noises in the audience aroused Clay's ever-ready suspicions, and he took his bowie knife from the carpetbag and put it under his belt. He did not provide himself with his pistols, however, an oversight he would soon regret.

When Turner finished speaking, Clay took the stand to introduce Curtis F. Burnam, who was to speak on behalf of his father. Cassius used the opportunity to belabor Turner for taking twice as much time as had been agreed upon. Vehemently he declared that Major Burnam's candidacy should receive an equal amount of time. But before Cassius could leave the platform, an unfortunate misunderstanding plunged the gathering into a confused melee.

There was a young lawyer present who had an entirely extraneous argument with Cassius Clay. Richard Runyon had been a member of the state legislature against which Clay had brought the charge of burning state school bonds. Earlier, Runyon had told Clay that the contention was not true and that he wanted a chance to tell his version. When Clay introduced young Burnam and berated Turner at Foxtown, Runyon had been in the rear of the crowd and had not heard Clay's argument. From Clay's vehemence, however, Runyon imagined that Cassius was repeating his charge about the school bonds. Running to the platform, the lawyer asked Clay if he had mentioned the bonds, intending to answer him if he had. Cassius, already bristling at the supposed slight by Turner, suspected that Runyon introduced the irrelevant matter to pick a fight with him, and he had no intentions of backing down. Those bonds, he truculently responded, had been burnt. Runyon yelled that it was not true. "Yes, sir," Clay rasped, "you voted for the bill to burn [them] . . . ask your master here, whose tool you are," he went on, pointing to Turner, "if I state not the truth."

When Cassius made that accusation, Cyrus Turner, eldest son of the candidate, stepped out of the crowd. Like everybody else, he had not understood the cryptic argument about the school bonds, but he had heard Clay insult his father. "You are a damned liar," Cyrus told Cassius, and struck him in the face. Cassius then drew the bowie knife from beneath his traveling robe, but Cyrus grabbed his arm so that he could not use it. Clay's knife was wrested from him by an unknown third party, and he slugged Cyrus Turner with his fist, knocking him back into the crowd.

With Clay the victor in the first round, and everyone confused about the origins of the fight, the tensions of the campaign suddenly exploded into a free-for-all fight. Nearly twenty relatives of Turner were present, and some of them began to attack Clay. His friends joined the affray on his side. One of the Turners had a heavy stick with which he pounded Cassius upon the head and back. Someone plunged a knife into his right side, and the blade went deep into his chest. At the same time, another Turner held a six-barrelled revolver at Clay's head and pulled the trigger three times before a Clay partisan threw him under the speakers' platform. Fortunately for Cassius, though the caps burst, the powder did not ignite.

Thrashing and striking blindly at his antagonists, Cassius espied his knife in someone's hand. He grabbed the blade in his hand and twisted it until he had retrieved it. In the process he cut three of his fingers to the bone. Blinded by the blows, short of breath because of the chest wound, and nearly overcome with pain, Cassius retreated. As soon as he had recovered his sight, he saw Cyrus Turner, whose blow had started the fight. Knife in hand, Cassius ran at him. Turner fell trying to escape, and as he lay upon the ground Cassius struck him a mighty blow in the abdomen and cut out his intestines. Then the crowd closed in and stopped the violence. Clay, who thought himself mortally wounded, dramatically shouted that he had fallen in defense of popular liberties.

He and Turner were carried to a nearby house and put to bed to await the arrival of physicians, while outside another dozen combatants bandaged cuts and nursed bruises. Though the campaign had suddenly erupted into war, the violence had arisen out of a misunderstanding rather than a genuine issue.

As he lay in the Foxtown bed, exhausted and in intense pain, Cassius could hear Cyrus Turner groaning in an adjoining bedroom. The first examination of the two men indicated that Clay would die of his chest wound but that Turner would live. The outcome was exactly the opposite; aften ten hours of agony, Turner died. Though Cassius suffered greatly, and all of his family expected him to succumb, he eventually recovered. He carried the marks of the conflict to his grave, and his chest pain never left him, but he gradually regained his strength. Five weeks after the encounter he reported that he was slowly improving. His deeply religious mother, who nursed him, marvelled at his recovery. "When I think how often your life according to human apearance [*sic*] was almost taken," she told him, "I wonder for what purpose it is sustained." [17]

Cassius might well have asked the same question, for the "fatal rencontre" at Foxtown had an adverse effect upon the emancipationist cause in Kentucky and upon the Whigs who were associated with it. The fight not only put Clay out of the campaign, but it also aroused much opposition to emancipationists in general. Cyrus Turner became a martyr, Cassius Clay was confirmed as a "damned nigger agitator," and "respectable" emancipationists would not be associated with so distasteful an affair. The result was the complete defeat of the gradual emancipationists. In Madison County, Squire Turner and William Chenault were chosen as delegates to the convention, while Burnam, the emancipationist, received only 688 votes out of a total of over 3,700. But Madison voters were not alone in rejecting the antislavery platform. Not a single emancipation candidate won a seat in the constitutional convention. When the new constitution was written, therefore, the perpetualists were in command.

Breckinridge had been right. The emancipationists demanded only the admission that the citizens had a right to decide the matter of slavery, and when they lost, they lost even that mild request. Cassius charged that the new constitution was an "infamous" document because it held that the right of the slaveholder to his "slave and the increase" was higher than any human or divine law. As the slave party detested northern abolitionists for their "higher law" doctrines, Cassius ridiculed the Kentucky Constitution of 1850 as supreme irony.[18]

The overwhelming defeat in 1849 brought changes to the emancipation forces in Kentucky; it was also a shock to the Whig Party in the state. The moderates, many of whom were Whigs, withdrew altogether and were willing to accept the Democratic perpetualist program as the will of the majority. Even Robert J. Breckinridge regarded the elections as final. "Having proved myself faithful to my convictions," he explained, "I shall now prove myself faithful to the Commonwealth." But Cassius Clay had long concluded that loyalty to Kentucky's best interests demanded liberation of her slaves, and he determined to persist in his efforts. "Others fell by the wayside," he recalled later. "I went on to the end." Though cursed and threatened, he made plans for another campaign.[19]

As Clay once more undertook to prepare campaign strategy, he took as his objective the destruction of the Kentucky Whigs. "My attack was mostly on the Whig Party—bent on its ruin;" he said later, "for, in our State it comprised a large majority of the slave-holders." The Kentucky Emancipationists, he announced, would offer candidates for the state elections of 1851. While he did not anticipate victory, he expected the diversion to defeat the Whigs and at the same time to reveal the antislavery strength in the state. "I think I shall get from 5 to 10 thousand votes, which will be a very good nucleus for future action," he told William H. Seward.

Cassius' brother-in-law, J. Speed Smith, who had served as state senator from Madison County, feared that the campaign would cost Clay's life. "Cassius has determined to have an emancipation candidate for Gov. and also one for Lieut.-Gov.," he told Brutus. "I candidly believe, should Cassius be a candidate for either station, the probability to be great that he would be *killed* before the election." Smith added that other Kentucky politicians concurred in that opinion. "There is more . . . hatred felt towards him than he is aware of." Cassius was but an "emancipationist upon high principles," said Smith, who knew him well, "but he is called and classed and believed to be an *abolitionist.*" [20]

There was, however, a growing basis for the charge that Clay had allied with northern abolitionists. More and more he proclaimed that that struggle was national, and he exhorted northerners to political action. "I am canvassing the State," he explained to some Free-Soilers in Maine, "with a view of organizing an Anti-Slavery party in Kentucky." He declared that the evil required a national remedy. "Of course it cannot be an issue between North and South," he assured them, "but between the slaveholding aristocracy of the South, and the large shipping merchants and cotton dealers of the North, on the one hand, and the great non-slaveholding masses of the whole Union on the other."

Clay contended that the struggle involved competing interests, not conflicting sections. He urged a political union of the antislavery forces of both sections. "We must arouse ourselves at once, or we are lost. Such men as Webster and Dickenson [21] in the North are traitors to freedom; they must be put down," he decreed. "In the South, the masters must learn that Slave States are not made up only of masters and slaves —but that another class, the great white laboring masses, must begin to be estimated in political calculations." He claimed that a national antislavery party would soon appear. "I think I see the elements of a truly national liberty party brewing," he confided to William H. Seward. "The True Democracy must be the name, since the Whigs have become the guardians of slavery." [22]

While he recognized the national aspects of the conflict, Clay was more interested in the local problem, and he prepared anew to organize a party of "true progress" in Kentucky. Although he called a convention to nominate candidates for the Emancipation ticket, it never met. Before the convention date, Cassius announced himself as the party's gubernatorial candidate. He invited Dr. George D. Blakey of Logan County to run as candidate for lieutenant-governor, and the ticket was complete. In the 1849 elections Blakey had been an Emancipation candidate and had come within one hundred votes of victory in his county. As the Clay-Blakey combination met the approval of the minuscule party, there was no need for a convention.[23]

The Kentucky Emancipationists met a mixed reception when they announced their candidates. The Democrats, taking the attitude that Cassius was still a Whig, were overjoyed at the prospect of a split in the opposition ranks. "Seeds of dissolution of the Whig party now ripen into harvest," one Democrat declared. But while Democrats encouraged the new development, Whigs ridiculed Clay's candidacy. "He is so perfectly harmless," a Whig editor remarked, "that his pretentions can only be made respectable by violent opposition. . . . We do not intend to make a 'grizzly' out of a 'bug' bear." In their efforts to discredit Clay, Whig leaders also enlisted the aid of the moderate emancipationists who had served in the Frankfort convention of 1849. Several of them complied, and published repudiations of Clay's party. "We think all will oppose this ill-advised and indiscreet movement of a few hot-headed fanatics," one of them said. "We deeply regret to see Col. Clay thus ranging himself alongside of the vilest Disunionists of the North." [24]

Undaunted by the partisan comment, Clay made plans for an active campaign. His first concern was for a party press. In Louisville, John C. Vaughan continued to publish the *Examiner,* which Clay adopted as his party journal. But he wanted a paper in central Kentucky. He paid six hundred dollars to D. L. Elder, an itinerant printer, to publish a sheet in Lexington, backing the Emancipation ticket. After printing three issues of *The Progress of the Age,* however, Elder became frightened and fled. Unfortunately, he failed to refund Clay's money before he left the state. Despite Elder's duplicity, Clay continued his efforts to establish an Emancipation press in the Bluegrass. He offered one thousand dollars to anyone who would issue such a paper for three years, but he found no takers. He did locate another antislavery paper in Kentucky, however. In Newport, a Kentucky suburb of Cincinnati, William S. Bailey, a poor machinist, was publishing the *News,* which supported Clay and Blakey.[25]

With newspapers in Louisville and in Newport along the Ohio River, the Emancipation candidates had to be content. To reach the voters, they planned a strenuous speaking campaign. Clay's bold and tireless travel in the canvass of 1851 became a part of the Clay legend. He appeared in about eighty of Kentucky's one hundred counties, and Blakey spoke in the others. It was widely circulated that at such engagements Clay would place a Bible and a bowie knife before him as guarantors of his right to speak, urging any who would not respect the one to beware the other. On June 2 he began his tour at Paint Lick Meeting House in Garrard County, and he spoke nearly every day until August 1. Many times he was threatened, and sometimes, with characteristic obstinacy, he spoke to completely empty rooms, but he kept every appointment without trouble.[26]

In his campaign speeches Cassius said little that was new. He repeated his contention that the two older parties had narrowed down to a single platform and no longer offered a choice of principle; therefore, he suggested the Emancipation Party as the party of opposition. He reiterated his time-worn argument that the majority of Kentuckians had no economic interest in slavery. He claimed that seven out of eight whites—" 'the people,' in the language of the politicians— . . . have no interest in the ownership of these slaves. . . ." He repeated his rejection of the religious approach. "I am not here as a moralist," he told his hearers, "but as a politician;" he chose to regard slavery, he said, "merely as a matter of dollars and cents." He concluded with his familiar statistics purporting to show the monetary cost of slavery to the state. "I think I have made out my case," he proclaimed, "that slavery wars upon the interests of the non-slaveholders of the State, the great majority of the people, and therefore ought to be overthrown!" Yet he maintained his respect for legal processes. "I hold that . . . the law is omnipotent," he said, and appealed for votes to bring about emancipation. Cassius made a determined effort to convince non-slaveholders that gradual emancipation would improve their economic position. But they refused to heed his arguments, and for fear, apathy, or prejudice, would not vote against the slave-owners.[27]

When the campaign was over and the returns were in, however, Clay was pleased to learn that he had been successful in his war upon the Whigs. In an election which counted more than 100,000 votes, the Democratic candidate, Lazarus W. Powell, won with a slim margin of 850 votes. Clay received only 3,621 ballots, but in such a close election the Emancipationists held the balance and threw the victory to the Democrats. More significant than the few who voted for the Emancipation candidates were the thousands who did not vote at all. Clay claimed that they registered a silent vote of dissatisfaction with the old parties, for, as the defeated Whigs themselves quickly pointed out, the winners had made no real gain in the election. In 1848, Powell had received 57,397 votes, but had lost; in 1851, because so many from both parties did not vote, he won with only 54,613, a decrease of nearly 3,000. The causes of the Democratic victory, one Whig explained, were the "indifference of the Whigs, some disaffection in regards to the candidate, and the diversion made by Cassius M. Clay, whose vote would have made the Whigs win."

But Clay took full credit for the Whig defeat. Although he received but three per cent of the total vote, it had been important in the result. Years later he boasted that in the 1851 campaign he had destroyed the Whig organization. "Thus, and forever, fell the Whig Party in Kentucky," he said. Under that name it would not win another election in the state.[28]

The election also brought about a change in Clay's tactics. He was now convinced that victory would not come through home action alone. After 1851, therefore, he no longer participated in Kentucky politics as a candidate, but he maintained an active, though small, antislavery party in the state. He had come to regard the problem as national rather than sectional. "I regard the liberation of my own state as the main object of my life," he told his friend Chase. "At the same time I feel that aid of public sentiment in the North is necessary to success here, and would make any personal sacrifice to forward the great cause of liberty to all, North and South, for it is at last one cause." [29]

Northern compromisers were as inimical to his ambitions as were the southern aristocrats. The growing antislavery political movement in the free states became, accordingly, more important to Clay, and in the decade of the 1850's he became increasingly involved in it. Perhaps, in a national organization dedicated to the defeat of the slave power, he would attain his goals.

CHAPTER XI

CLAY BECOMES A REPUBLICAN

THE election of 1851, which marked the defeat of the Kentucky Whig Party, was a turning point for Cassius Clay. After years of constant agitation he was still far from his goal, and he abandoned further serious efforts to win power through state action. A new political development in the North offered him the possibility of victory over slaveholders on the national scene. In 1848 the Free-Soil Party had shown surprising strength by winning the balance of power in twelve states and enabling Taylor to win. Taking as its motto "Free Soil, Free Speech, and Free Men," the party platform declared that the western territories ought to be free. Its vote in 1848 convinced Clay that there was vitality in the Free-Soil program. After 1851, he participated in the northern movement, seeking to become the leader of an uncompromising national antislavery party.

Clay wasted no time in beginning his career as a northern Free-Soiler. In September, 1851, less than a month after the gubernatorial election, he met with leading northern antislavery politicians in a national convention at Cleveland. "I differ with your friends in many subordinate matters," he told Ohio abolitionist Joshua R. Giddings, "but I am willing to merge with them in the great question." The disappointing results in Kentucky and the growth of Free-Soil sentiment helped Clay to overcome his scruples, and he took his first step toward a political alliance with northerners. He moved cautiously into political abolitionism, trying to placate his Ohio friend Salmon P. Chase and his New York political ally William H. Seward. "You lead me into unknown seas," he joked to Chase. "Who shall pilot me out once more?" But Clay resisted any effort to make him bear the burden of a hopeless campaign. "Some have talked of nominating me vice president at Cleveland . . . ," he told Chase. "I think as I have been in the post of danger for long years, I should be allowed to wait the tide of *success*, or hold the *first* place in a 'forlorn hope.' "[1]

He was not overly disappointed, therefore, when a nominating convention of the Free-Soil Party named Senator John P. Hale of New Hampshire presidential candidate for 1852, and decided upon George W. Julian of Indiana for vice-president. Clay spoke in Kentucky on behalf of the Free-Soil candidates, and he escorted Julian on a tour through the Ohio River section of the state. Although they visited only border areas where they expected support, the response was poor. At one place the local rowdies, "low-bred tools," Cassius called them, soiled the courtroom with excrement, and the campaigners were overpowered by the odor. Julian, who had not lost his sense of humor, surveyed the smelly scene with disgust and said, "Well, we should not complain of the Slave-Power; for this is the *strongest* argument they could have presented! . . . The remarkable thing was that they did not mob us."

Nor did Kentuckians cast their votes for Free-Soilers. The party polled only 266 votes in the state. Clay's party had ceased to exist as an effective organization, but he was undismayed. He expected his reward from another source, and he renewed his efforts to attract a northern following. "The cause of emancipation advances only with agitation," he counselled. "Let that cease, and despotism is complete."[2]

To arouse northern voters to the evils of the slave system was Clay's first task, but after the exciting debates of 1850 it was difficult to maintain interest in the antislavery crusade. Many northern citizens despised the erratic, inflammatory abolitionist agitators, and preferred to consider the matter settled. The problem Clay, in common with other Free-Soilers, faced was keeping open the discussion of slavery. To attract audiences, party leaders resorted to the unusual: escaped slave Frederick Douglass, an intelligent and forceful speaker; and reformed slave-owner, Cassius M. Clay, a popular lecturer. That northerners could be aroused to the evils of slavery was indicated by the spectacular success of Harriet Beecher Stowe's celebrated novel, *Uncle Tom's Cabin.* "Agitation then ought not to cease," Clay advised. " 'Uncle Tom' proves that there is vitality in it!" He commended Mrs. Stowe's accuracy, and added, "For one case of 'Legreeism,' I'll show you a dozen of infinitely exceeding horror!" Clay, who proclaimed himself an expert on things southern, found that the sadistic, rather than the statistic, influenced his hearers.[3]

Unusual speakers to attract the curious, and sentimental novels to affect the imaginative, helped awaken interest in the threat of slavery, but an even greater service came from Stephen A. Douglas, a young, ambitious senator from Illinois. The "Little Giant" dreamed of a transcontinental railway whose eastern terminal would be his home town, Chicago—and for this reason, among others, he was concerned about western territorial organization.

The route west from Lake Michigan ran through a vast unorganized area still subject to Indian raids. To make his plan feasible, Douglas needed a territorial government for that section of the Louisiana Purchase, and needed also the protection of the United States Army. The canny senator realized that his southern colleagues, who had railway schemes of their own, would resist an agreement to aid a competitive route unless they received some concession. Douglas suggested, therefore, that the Compromise of 1850, which had permitted New Mexico and Utah to decide for themselves whether they should become free or slave territories, had in fact repealed the Missouri Compromise of 1820. The Kansas and Nebraska territories, although they were in that part of the Louisiana Purchase closed to slavery by the 1820 agreement, ought to have the same right to choose for themselves. The Illinois senator thus appealed to the popular sovereignty sentiments of westerners as well as to the proslavery prejudices of southerners.

Because it posed the threat of a new expansion of slavery, the scheme aroused an immediate reaction from the antislavery forces. The repeal of the "time-honored" Missouri Compromise, Clay recalled, aroused an "alarm and indignation in the Nation which was never before witnessed." It demonstrated, he said, his oft-repeated contention that "there could be no compromise between Liberty and Slavery." Clay toured the Midwest to take advantage of the popular discontent and to assist in Douglas' defeat. He invited those "in favor of Republicanism and against the Divine right of Kings" to join him in opposing the Nebraska bill. "It is not a duty to form a crusade against the South for moral reformation, so long as Slavery is confined to its own ground," he said; "but when it proposes . . . to become aggressive, to go into free territories, then I will stigmatize it." Clay had taken another broad step toward the North; he had adopted the Free-Soil platform. He still did not like that name, however. "I have not advised a change of name, though I feel very indifferent about it," he told Chase. "The name of Republican adopted in several states is significant." [4]

Clay's journey through the Midwest was but the prelude to an intensive speaking tour which carried him all over the North. As a professional lecturer, he received fifty dollars a speech for his anti-Nebraska views. "In my humble way of nightly lectures, mostly on the despotism of slavery, I trust

I am doing good service in the common cause," he explained. "My audiences are overflowing and enthusiastic." He employed bitter ridicule and sarcastic taunts to arouse his hearers to action. The system of slavery, he repeated, destroyed not only the "liberty of the colored, but of the white population; and not only of the white population of the Slave States, but also of the Free States." Clay bemoaned the condition of civil liberties in the South and then sarcastically demanded, "How much better off are the people of the North? They have not the right of speech in Congress, nor of petition." Using arguments he had stereotyped in the Kentucky hustings, Clay asserted that northerners were victims, along with southerners, of an antirepublican despotism. Yankees were "bloodhounds for the South, on all the territory where their four-footed bloodhounds dared not venture." The political lesson he taught was clear. "I can not see how Northern freemen can unite themselves with the Democracy," he lectured. "The Democracy has always had a stronghold in South Carolina, the least Democratic of all the States." To further his own ambitions, and to bring about certain public benefits he confidently anticipated, Clay sought to discredit a national party, the Democratic, in order to enhance a sectional party, the Free-Soil–Republican.[5]

While he was trying to arouse northerners to resist the slave-owners, in Kentucky he set them an example by continuing his belligerence. In the summer of 1855, a Lincoln County group suppressed Clay's colleague, John G. Fee. Fee, a tiny, wizened man with straggling whiskers, was a native Kentuckian who had graduated from Cincinnati's Lane Seminary, where he had imbibed an antislavery religion. Upon his return to the state he became minister of churches which forbade membership to slave-owners. Clay invited Fee to Madison County and provided land upon which the little man established a "higher law" church and a school which became

Berea College. Fee denied the legality of slavery, and on religious grounds refused to recognize it. Though his preachments aroused fierce resentment, Fee was a non-resister.[6]

Clay was no pacifist, but boasted that he was a "fighting Christian." When an audience prevented Fee from speaking in Lincoln County, Cassius promptly assembled a company of armed ruffians and traveled to the scene of the dispute. Dramatically he announced that he would speak there, freely if possible, "*by force*" if necessary. He argued that the principle of free speech was at issue, and he intended to defend it. Guarded by the private army, Clay finished his address without trouble and proclaimed that he had once more established freedom of speech in Kentucky. Repelled by his show of force, however, his critics alleged that he had trampled upon as much freedom as he had rescued. His gang, according to a report from Lincoln County, was "armed to the teeth" with firearms and butcher knives, and was as much a mob as those he opposed. But Cassius seemed unaware that his strong-arm methods rendered him unpopular, and he used the incident to prove his contention that resistance brought results. "The mob party," he said (speaking of the slaveholders), "quailed before manly resistance." But forceful opposition in slave states was not enough, he declared. "If the North gives way on the Kansas conquest, I and our party will be destroyed," he predicted; "if they keep up a manly opposition, I think we will come out victorious." [7]

Clay had become a Republican, and he entrusted his ambitions to the northern party. "I am cheered that I find myself in sympathy with the great minds and heroic hearts of the Nation," he said. "All hail, the North!—all hail, the Republican Party." Proclaiming a new political allegiance, Clay worked to make the party an antislavery instrument. "Our only salvation, because the only true repentance," he exhorted, "is in making the overthrow of slavery our dominant idea."

The party must face slavery, he suggested, "not compromisingly, . . . but with a . . . fanaticism of will." Clay demanded that his northern allies stand firm against his personal enemies, as he had done, regardless of the consequences. "I am for no Union without Liberty," he asserted, "if need be through dissolution and war." Already he had accepted the idea that war was inevitable. "We shall not have a peaceful triumph," he advised party members. And to a friend in Kansas he said, "You will have to fight again or be subjected. Mark what I tell you. Unless you are prepared to repel force from slaveholders, you will never have peace. I have tried them for twenty years. They have no magnanimity, no remorse, no mercy!" [8]

Clay's implacable opposition to slavery influenced the organization of the new Republican Party. In February, 1856, in a special convention at Pittsburgh, the national party was born. Some of the organizers, led by Missouri's Francis P. Blair, counseled moderation and a spirit of compromise toward slaveholders. Governor Kinsley S. Bingham of Michigan was the leader of the irreconcilables. To counter Blair's testimony, Bingham read a letter from Cassius Clay. After hearing Clay's words, the delegates adopted a stronger antislavery resolution than they had planned. George W. Julian, one of the delegates, said that the "impassioned and powerful arraignment of slavery by a Southern man [performed an] excellent service in guiding and inspiring the great party." Clay's efforts to enlist the aid of a northern party in his war with slave-owners was nearing success, and influential Republicans pledged themselves to him. "I understand Vaughan of the Chi: Tribune is out for Fremont," he told Chase. "If so, he has violated his pledge to me, to bring out my name . . . backed by a concerted movement from Illinois!" But he could still claim that he had never canvassed support. "With regard to myself, I have not been a candidate for either office," he said to Chase after the Republican nominating convention of 1856. "I have never asked a man on earth to go for me; and I don't know that I ever shall, unless I clearly change my mind." [9]

Clay was encouraged at the prospects for the new party, but his private affairs were not so happy. The prewar decade marked the nadir of his financial difficulties. In an effort to raise money he had resorted to drastic measures, such as speculation in pork and land. Having been unsuccessful with his own finances, in 1854 he set himself up to manage other people's money. On November 8, he announced the opening of the firm of Cassius M. Clay and Company, Bankers, at Number 59, Third Street, Cincinnati. But on the very day he opened his business, a severe bank crisis forced the leading banks of the city to suspend specie payments. Once more Clay's bad luck had followed him. The depression and the ensuing scarcity of specie doomed his enterprise, and three months after its opening the firm dissolved.[10]

And his financial troubles continued to mount. A sharp decline in prices upset his speculative business in pork, and in 1856 he announced his financial failure. He made an assignment of his entire assets, which his assignees offered at a public sale to settle his accounts. "I am sorry to tell you that I am *broken*, and have mortgaged my effects," he sadly informed a friend. "I am low in spirits, and we are arranging our furniture for a sale." On April 15, 1856, he held an auction at White Hall and watched his prized possessions go under the hammer. There were large French mirrors; a collection of fine paintings; several pieces of marble statuary; assorted musical instruments; and a library of over five hundred volumes, including works of history, biography, law, philosophy, travel, and *belles-lettres*. Clay's mother and brother bought much of the property and then allowed him to use it, so the sale transferred his debt to his family.[11]

As his household furnishings indicated, Clay lived luxuriously. Accustomed to the life of the southern landed aristocracy, he was a *bon vivant* and a patron of the arts. He enjoyed a hunting or fishing trip and delighted in good-fellowship and conviviality. He was well acquainted with many outstanding literary and artistic personalities of his day—men like the visiting Norwegian musician, Ole Bull, the New England essayist and lecturer, Ralph Waldo Emerson, and the noted educator, Horace Mann. Clay was also interested in Kentucky artists; he subsidized Joel T. Hart, who became the state's most famous sculptor.

Along with his outside interests, Clay had a growing family to care for. He and Mary Jane had added to their family: in the early 1850's, two daughters, Laura and Annie, were born. He was away from home much of the time on lecture tours and business trips, and he complained that his children did not really know him. His financial collapse in 1856 placed severe strains upon his family ties. When a friend offered to send a photograph of Clay, the despondent man replied, "Mrs. C. is not much in love with my face *now*, so you had better not perhaps send the portrait." [12]

It was with his wife's family, however, that he had the most serious quarrel. He was angry that none of the Warfields had come to assist him in his monetary difficulties. "Some friends deserted me, others stood up nobly," he told Chase. "My father in law, who is very wealthy, and who has never made me any advances, meanly left me, as he supposed, to ruin and the streets! He never came near me on sale day. Nor will I ever care to see him again!" [13]

Clay lived a full life, but it was as political agitator that he was most widely known. He allowed neither his personal affairs nor his family problems to keep him out of Republican Party activities. After 1856, his ambitions for public office became more obvious and more insistent. In that year he indirectly sought a nomination at the national Republican convention. Before party members assembled he advised them that they should choose candidates who would not compromise the slavery issue. "Let that candidate," he said, "whether [Thomas Hart] Benton, [William H.] Seward, or [John P.] Hale, or any other good citizen, be chosen without regard to his locality in a Free or Slave State." Clay's activities indicated that he considered himself eminently available, but once more he went unrewarded. In the selection of a vice-presidential candidate, in which an unknown from Illinois, Abraham Lincoln, showed surprising strength, Clay received a few complimentary votes, but the nominations went to John C. Fremont and W. L. Dayton.[14]

Clay, like his friends Seward and Chase, swallowed his disappointment at not being nominated. Although he was not a candidate, he played an enthusiastic part in the campaign. In Kentucky his skeleton party put up a slate of electors and made the motions of conducting a campaign, but Clay was more interested in keeping alive the belief that he had a Kentucky Republican following than in any real anticipation of victory.

As in 1852, Clay's major effort was in the free states, where he made numerous appearances. Again he tried to insult his hearers into antisouthern action. Commenting upon the caning of Senator Charles Sumner by a young South Carolina congressman, Clay alleged that southerners made such attacks because "they believed the North lacked courage." He said that when he spoke in Kentucky about the growing antisouthern sentiment in the North, his neighbors always replied, "We don't care a d——d, we have bought your doughfaces and can buy them again." His main theme, as it had been earlier, was to urge northerners to stop "acquiescing in [southern] aggressions," and to resist slave expansion.[15]

Great crowds gathered to hear Clay's militant antislavery speeches, and many listeners professed that he had converted them. "Clay's arguments, based upon personal experiences as they were," one hearer said, "had a strong effect in strengthening the determination that Slavery's curse shall not be extended over free soil." And in an Indiana community an old man staged a convincing bit of political histrionics. Wearing a Buchanan badge, he attended a Clay rally. After Clay's fiery speech he tore the button from his hat and loudly recanted his past sins. Republican leaders missed no chance to dramatize popular resentment against the extension of slavery into the territories. The enthusiastic reception encouraged Clay. "Things look well everywhere; we will *win* now!" he exulted to Chase. "Indiana certain." [16]

To Clay, however, the most important feature of the 1856 campaign, as he reviewed it in later years, was his first encounter with Abraham Lincoln, the man who would mean so much to him and to the nation in the coming crisis. During the campaign Clay spoke at an outdoor political rally in Springfield, Illinois. As he spoke, Lincoln sat under a nearby tree, whittling thoughtfully on a stick as he listened to the speech. When Clay finished, Lincoln slowly arose and commended him.

Since Clay's name had become a household word after the suppression of *The True American*, Lincoln had followed his career through the Lexington newspapers which Mrs. Lincoln received. This was the first meeting of the two men, but it was not to be the last. Shortly after the Springfield rally, Clay met Lincoln on a train as they journeyed to speaking engagements. As he remembered it, Clay made use of the opportunity to "convert" the tall Illinois man to his antislavery opinions. Lincoln listened to Clay without saying anything himself. Then, when Clay had finished, Lincoln, perhaps with comic gravity, replied, "Yes, I always thought, Mr. Clay, that the man who made the corn should eat the corn." Enthusiastically, Clay congratulated himself upon a great victory. But as events were to prove, Lincoln was far too astute a man to follow an erratic extremist like Clay.[17]

Clay spoke effectively in many midwestern towns, but he saved his most polished oratory for an audience in New York City. Addressing the Young Men's Republican Central Committee in the Tabernacle, he made a determined bid for the support of that influential group. As a background for his claim to high office, Clay reviewed his entire war upon slavery. In his customary garb of dark suit with brass buttons and with a face serious under his dark hair (now showing streaks of gray), he was a memorable figure as he reiterated his familiar contention that slavery prevented southern economic expansion. Once again he was a spokesman for southern manufactures, but it was the measure of his failure that he now preached to New Yorkers.

There were, in the South, ample resources of power and minerals for an industrial economy. "We have taken Man, and subjected him to our will," he told his metropolitan audience; "you have . . . seized upon the elements—upon steam, upon water-power—upon chemistry, upon electricity . . . and made them your omnipotent slaves." The same results would be achieved in the South, he said, but for the emphasis upon the slave-worked agriculture. In western Virginia and in eastern Kentucky, "coal, and iron, and marble, and other minerals of unequalled value," were readily available, but remained untouched. "There is the Blue Ridge penetrating the clouds and pouring down perennial streams of water-power," Clay asserted; but he declared that it was wasted in a section "without manufactures sufficient to clothe her half-naked slaves." If southerners would harness that power, the whole of southern economy would prosper, he promised. "Without manufactures and mining there is no commerce, and with the finest harbors in the world, there is not a ship upon her stocks, nor a sail unfurled." [18]

Yet in the face of such arguments, he continued, southerners would not forego their preference for cotton. In the South, whenever he presented his plea for manufactures, he said, his hearers always "cut short the argument by lustily crying out at the top of their voices: 'O! cotton is King!' On the contrary," Clay shouted, "I proclaim that—grass is King!" As evidence, he compared the value of the cotton crop with that of hay, as given in the census returns, and reported his finding that hay was of much greater value.

The argument that cotton brought a monetary return because it entered foreign trade Clay also denied. "If you were to blot out the whole foreign trade in cotton, the country . . . would be much the gainer, in domestic industry, in home manufactures, home labor, and a home market for ourselves," he asserted. For twenty years he had explained southern technological backwardness in terms of its slave labor system. Now, as a Republican, he tried to persuade northern voters to share his enmity toward the defenders of slavery.

Sarcastically he described the poor state of southern industry: "In vain do men go to Nashville, and to Knoxville, and to Memphis, and Charleston, in their annual farce of southern commercial conventions to build up Southern commerce, and to break down the abolition cities Philadelphia, Boston, and New York. The orator rises upon a northern-made carpet; clothed cap-a-pie in northern fabrics, and offers his resolutions written upon northern paper with a northern pen, and returns to his home on a northern car; and being killed, is put into a northern shroud, and buried in a northern coffin, and his funeral preached from a text from a northern hymn-book, set to northern music. And they resolve and resolve, and forthwith there's not another ton of shipping built, or added to the manufactures of the South, and yet these men are not fools! They never invite such men as I to their conventions, because I would tell them that slavery was the cause of their poverty, and that it is free labor which they need. . . ." [19]

Clay's New York speech brought high praise and won him several friends who campaigned for him prior to the 1860 convention, but it had little effect upon the current election.

Buchanan and the Democrats won the South as well as the northern states of Illinois, Indiana (despite Clay's confidence there), and Pennsylvania, which would accordingly become, in 1860, the prime targets of the Republicans. But though he lost, Fremont did surprisingly well. He polled 1,300,000 votes and very nearly united the North and West against the South. The election revealed party alignments which approached a division along the Mason-Dixon line and thus foreshadowed the election of 1860. Kentucky cast her first Democratic electoral votes since the "recognition of parties," demonstrating southern sectional loyalties. Clay's Kentucky Republican Party registered no gain at all. In a total vote of over 100,000, Republican electors received only 300 votes. To Cassius Clay, the election of 1856 showed clearly that he must look northward for the fulfillment of his ambitions.[20]

Clay's party did not offer candidates in 1858; and in the following year, although he made no campaign, Clay received one vote for governor. That, his opponents chortled, was the measure of the Kentucky Republican Party. Clay angrily responded that the party might not be numerous enough to meet the "wide vision" of its critics, "but it is large enough to stand by all its convictions, and defend its rights, whenever with speech, the pen, or the *sword*, it is attacked by despots!" [21]

Although his Kentucky following had withered away under the glare of the sectional controversy, Clay's fierce ambition would not let him admit defeat. In the years before 1860, he dropped hints about the "mistake" of having no southern man on the ticket, and about his own availability. "There was a great error in having both men North of Mason's and Dixon's line—it gave our enemies the vantage ground just where we were weakest," he counseled Chase. "I don't say this on my own account, because I did not desire a nomination, when success was doubtful." But he anticipated victory in the next election, as he told Seward: "I think that it will be the grossest mismanagement on our part if we do not carry our candidate with ease in '60, whoever he may be." For that reason, he earnestly desired a place on the ticket. "We must have one candidate of the two, this next canvass, *South* of the line."

He expected no support from the slave states. "We can hardly hope for an electoral vote in the South," he said in an 1858 estimate to Seward. "We might get Maryland or Kentucky in certain emergencies," he went on, "but in such case the slavery question—the only question outside of the 'ins and outs', would be ignored." Twenty years of fruitless agitation had convinced Clay that political victory over slavery was unlikely in the South. "The battle is to be fought by the free states: and their views should be kept always foremost." Hoping for political reward from the North, he continued to make suggestions about the coming nominations. "Our policy then seems to be to sustain our old platform rigidly, put up a representative man in '60, and fight upon *our principles uncommitted to any compromise*," he advised Chase.[22]

As the crucial year approached, Clay renewed his efforts to obtain a nomination. His friends worked to secure him contacts and speaking dates and wrote to influential party men for him. Clay began his campaign in January, 1860, with a bold speech from the steps of the State Capitol in Frankfort. Because the building was closed to him, he spoke outside in darkness and drizzling rain to several hundred patient listeners. Though he spoke under unusual handicaps, the speech was one of his best. For three hours he held his audience through a complete exposition of all current political issues, from the recent uproar over John Brown's raid into Virginia to a list of seventeen charges against the Democratic Party. It was an able defense of Republican principles delivered in the heart of a slave state, and it attracted wide interest. Over two hundred thousand copies of it were distributed as a campaign document.[23]

A month later, on February 15, Clay spoke in Cooper Union

and became more obviously a candidate. The secretary of the New York Young Men's Republican Union organized a series of meetings which would bring political aspirants before metropolitan voters. "Confidentially, I say frankly," he told Cassius, "the whole thing is intended by me as a Clay demonstration." He gave Clay a favored spot in the program and offered him an unusual opportunity to win support. When Clay stood before the crowded New York hall, he spoke with more candor than usual. The party had no intention of harming slavery in the states, he told the slaveless New Yorkers. "The North simply intends to take the reins of Government into their own hands." Republicans opposed slave expansion, he said, not for any moral sentiment, but for the more practical reason that they "knew they could sell more to Free States than they could to Slave States."

Clay repeated his contention that the situation required a firm, unyielding leader who understood southerners, and he frankly declared that he was the man. He had fought southerners for twenty years, he boasted. "They can't drive me out, gentlemen . . . it is not safe to put down 'Cash' Clay!" Northerners, however, were afraid of the South and had surrendered too easily. "Put me at the head of the United States," he challenged, "and I will whip them." Clay pretended surprise at the applause which followed. "Why, you seem to be really in earnest about it," he said. At that, a Clay supporter yelled, "All in favor of Mr. Clay being the nominee for the next Presidency, please say 'Aye,'" and the hall resounded with shouts. But Clay would learn that New Yorkers would have little voice in the nomination. The next speaker on the Cooper Institute program was a tall man from Illinois, Abraham Lincoln.[24]

Quite pleased with the reception in New York and his seemingly bright future, Clay returned to Kentucky, where he again openly avowed his ambitions. On a Lexington street he met Leslie Combs, an old-time Whig and one of Henry Clay's chief lieutenants. Pounding Combs on the back, Clay told him, "You can make me President, if you will." Combs, an influential Kentucky Unionist, promptly responded that he was "dead against" the Democrats who currently ruled Kentucky with a "rod of iron," but that he was "equally opposed to abolitionists—if you are one." Clay assured Combs that he had always favored legal means of emancipation, and he offered to send Combs a copy of the Frankfort speech which explained his position.[25]

Encouraged by many tokens of his popularity, Clay made an effort to secure the support of New York's political manipulator, Thurlow Weed. He stated that William H. Seward was his first choice for the nomination, but if "*newcomers*" were to enter the balloting, then he would enter the race himself. To Weed he wrote:

> There is a widespread and increasing belief that in that event I will be chosen for these reasons:
>
> 1. I am a Southern man, and the cry about sectionalism will be silenced.
>
> 2. I am a tariff man: and Pa. must be consulted in that.
>
> 3. I am *popular* with the Germans everywhere, and not offensive to the Americans [by which he meant citizens of other national origins].
>
> 4. I have served the party longer than any other man without contemporary reward as others have had.
>
> 5. There are elements in my history which will arouse popular enthusiasm and *insure without fail success.*
>
> 6. That I will form a Southern wing to the party which is necessary to a safe administration of the government, and thus put down all hopes of *disunion.*

He could list other attributes, the ambitious Clay went on, but these, he reasoned, should be sufficient. "I think I am the *second* choice of all the 'old line' candidates' friends," he concluded.[26]

Despite Clay's unblushing effort to secure the support of the party whip, the delegates at the nominating convention overlooked him. When a combination of minority groups united to defeat Seward in the Chicago session, they also ruined Clay's chances for second place. Abraham Lincoln, the victorious nominee, was, like Clay, a native Kentuckian with a Whig background, and the politicians needed a more balanced ticket. For their vice-presidential candidate, therefore, they chose a man from New England with a Democratic background. On the first ballot for the vice-presidential nomination, Hannibal Hamlin of Maine received 194 votes and Clay got 101½. But on the second vote Hamlin won enough of the favorite-son votes to secure the nomination. As soon as the outcome was evident, George D. Blakey, Clay's lieutenant on the floor, arose and moved that the nomination be made unanimous. As a consolation to Clay, the delegates dutifully offered him three cheers.[27]

Clay's aspirations were once again frustrated, and his failure was not solely due to the turn of the political wheel. Because he had identified himself as the spokesman for southern industry, his appeal was severely localized. In 1856, in his New York Tabernacle speech, his main theme had been the industrialization of the South through the exploitation of mountain resources and the defeat of slavepower. Again, in 1860, at the Cooper Institute, the only program he had to offer was firm leadership to "whip" the southerners. So closely had he associated himself with the needs of a backward region that he had little to offer as a national candidate. He had no positive program for residents of free states other than resistance to slave expansion. His militant strictures against southerners made him unacceptable to moderates on either side of the Ohio River. The weakness of Cassius Clay's program was that it was regional, not national; even worse, his belligerent exhortations urging war made him a liability to the party. Clay wanted to make the Republican organization what southern fire-eaters termed it: a declaration of war against the South.

Though Clay's defeat disappointed him, he hid his grief. He congratulated Lincoln upon the nomination, but added that "the mistake of putting no Southern man on the ticket will weaken our efforts in the Cause here immensely." Cassius Clay had no intention of abandoning his hopes; he was too ambitious to sulk in his tent. "We must with a good grace submit," he told Lincoln. Clay's strength at Chicago had not been sufficient to win a nomination; nevertheless, he had polled a respectable vote. If he participated in a victorious campaign, he might yet attain high office.[28]

CHAPTER XII

REWARD FOR AN AMBITIOUS MAN

"WELL, you have 'cleared us all out!'" Cassius Clay exclaimed to Abraham Lincoln after the Chicago Republican Convention. To hide his disappointment, Clay pretended that he had not really wanted the nomination. "I did not at all press my claim *now*," he said, and he denied any desire for the vice-presidency. "I may say," he confided to his friend, Ohio's Senator Salmon P. Chase, "that I was indifferent as to being the 2nd candidate." And to New York's unhappy Seward, Clay sang a similar tune: "I cared nothing for the vice Presidency, it being a 'switch off' so far as future promotion is concerned." Despite his disavowals of political aspirations in 1860, Clay made no effort to hide his hope for future promotion. "Next time my friends will press me for the first post with earnest hopes of triumph," he said. Until then, he needed a place of prominence which would afford him an opportunity to demonstrate his abilities and win public support. With that objective in mind, Clay would use any pretext which might influence Republican leaders.[1]

From the first, he offered the plea that his long years of service to the party entitled him to a place of prominence in it. He was a charter member of the Republican Party, he pointed out. Indeed he had served the cause long before there was a party bearing that name, and he often grumbled to party leaders that he had labored longer without reward than any of them. But once more he offered his services for the campaign. In accepting his help, Lincoln responded in words which the overeager Clay interpreted as the promise of a place in the cabinet: "I shall in the canvass, and especially afterwards, if the results shall devolve the administration upon me, need the support of all the talent, popularity, and courage, North and South, which is in the party. . . ." Office-hungry, Clay saw in that routine remark the promise that he would not be forgotten should the party win the election.[2]

But before any Republican could receive his share of the spoils, the party must win in the forthcoming contest. The problem was simply stated: Republicans must win over to their cause enough Democrats in key northern states, and must continue to hold their 1856 gains. Kentucky was not essential to the victory, so Clay's campaign in his own state was perfunctory. There he characterized Lincoln as a local product—a "one-gallows barefoot boy."

Clay did not remain long among his neighbors. After opening his speaking tour in Louisville, he spent most of his time north of the Ohio River. Indiana, one of the few non-slave states which had resisted the Republican tide four years earlier, was now a prime target of the party. A local politician invited Clay into the southern part of the Hoosier State to assure voters of Republican conservatism. "The people in that portion of the state have been taught by the Democracy that you and every prominent Republican desires the immediate abolition of slavery and the elevation of the negro to an immediate social and political equality with the whites!" he told Clay. Clay's task was to counter that allegation, which, to a man of his marked prejudices against the Negro, required no subterfuge. It was one of the rare occasions when he was summoned to defend a conservative position.[3]

In Indiana, Clay's rallies included typical campaign trappings: cannons, torch-parades, wigwams, split logs. Republican campaign managers understood full well how to make the most of an attraction like Cassius Clay. When he appeared, the "Lincoln and Hamlin" banners shared attention with others declaring "Clay, the Champion of Free Speech, Welcome!" Along with the side-show atmosphere went Clay's two-hour-long speeches, happily spiced with witty extemporaneous banter. He was serious, however, when he denied any sympathy with radical abolitionism, and he whitewashed the party's position on slavery. Tirelessly he worked to convert skeptical Democrats into loyal Republicans. His hair now gray and his pale face beginning to reveal a florid tint, Clay grew weary long before his one hundred personal appearances were over. He was glad when his schedule was completed and he could return to White Hall to recuperate.[4]

The Republican victory pleased Clay, and because Indiana did vote for Lincoln, he expected his labors to bring their reward. But as weeks passed without word from Springfield, he became concerned. Perhaps the Lincoln coterie did not fully appreciate his years of unrequited struggle. "For twenty years I have been in *exile* for principle's sake," he reminded his friend Chase. "Now when those to whose magnanimity I trusted my all, have come into power, they propose to ignore me!"

Quietly he prompted his friends to speak up on his behalf, and many of them wrote to Lincoln about Clay's heroic services. An Illinois voter suggested that Clay should receive the war secretaryship because of his understanding of the southern temper, and a Kentucky mountaineer praised Clay's unending struggle against the proslavery forces. "Thar is no man that done more and sacrfised more than Mr. Clay for the Republican Party," he scrawled. "Thar is a grate many of our Enemies would rejoice if Mr. Clay is passed by and they will bring everything to bair aggainst him in their power."[5]

As the mountaineer suspected, Clay's enemies were already at work to foil his ambitions. Kentucky conservatives, who had opposed Clay's unconventional career from the first, were frightened by the prospect of a national administration with him in a prominent position. On November 16, their spokesman, the aging Daniel Breck, journeyed to Springfield to consult with the President-elect. Breck was an old-line Whig with whom Cassius had worked in 1840, and he had married an aunt of Mrs. Lincoln; he was, therefore, a well-chosen representative of Kentucky moderates.

The old man advised the President-elect that, to save the Union in the mounting wave of secession-fever, he should surround himself with conservative men. One or more of his cabinet, Breck insisted, should be southern non-Republicans. If Lincoln followed such a course, the Kentuckian assured him, then his native state would remain in the Union. But, Breck warned, if the President-elect should choose an "obnoxious" man, like Cassius M. Clay, Kentuckians would regard it as a "declaration of war against the state." [6]

Patiently, Lincoln heard Breck for two hours before making a direct response. Then, when the old Judge had finished his plea for a cosmopolitan cabinet of moderate northerners and proslavery southerners as the price of union, Lincoln slowly replied, "Does any man think that I will take to my bosom an enemy?" Though he thus rejected Breck's request for putting anti-Republicans in the cabinet, the President-elect accepted the Kentuckian's estimate of public sentiment concerning Cassius M. Clay. Clay's consistent pugnacity had associated him with the radical wing of the party and rendered him unacceptable as a member of the administration. At a time when the South was seething with confused fears, Clay was not the southern man Lincoln wanted for his cabinet.[7]

Unaware that his forthright militancy had doomed his aspirations, Clay continued to issue warlike threats to defeated southerners as though he were the party's policy-maker. He acted as if, being a southerner himself, he had to be more rabidly antisouthern than were northerners, if he would win their confidence and their patronage. Clay declared that it was the duty of northerners to coerce the seceding cotton states back into the Union. "Every man of sense sees that civil war would be better than eternal war which would be the result of a divided nation," he thundered. But though he demonstrated his devotion to the Union by stern denunciation of southern political activity, still Clay received no offer from Lincoln. The vexing silence from Springfield perturbed him, and his plaints became querulous.[8]

Finally, on January 10, 1861, Clay could stand the strain no longer. Swallowing his pride, he wrote a long letter to the President-elect, angrily demanding consideration. He reviewed his extended struggle against slavery and against the slave party and claimed a cabinet post on the basis of it. "In the success of the party, which you represent, I did feel that my long though humble services, did entitle me to a portion of the controlling interest in the administration of its destiny," he informed Lincoln. To refute Daniel Breck's charges, Clay protested that no other southern Republican would give more confidence to the administration. He lost his temper when he censured Lincoln for readily accepting Breck's analysis of Kentucky sentiment. "If I was fit for the place—then I should have been judged only by my own merits and your *individual feelings*, nothing should have been yielded to clamor," he urged, "nothing to a *false policy* . . . which offers the rewards of triumph as a premium to meanness of spirit, indifference to principle, and personal cowardice!" Then, more calmly, Clay dropped the hint that he would accept a foreign ministry, though only to England or France.[9]

But his frantic efforts were to no avail. Because of Clay's taint of radical abolitionism, Lincoln had already decided against appointing him to the cabinet. Moreover, Kentucky was essential to Lincoln's Union-saving schemes, and he would take no step which might alienate her people. In the opinions of those who directed the newly elected party, Clay was too dangerous. His old friend, Missouri Congressman James S. Rollins—who had been best man in his wedding, and his representative in the difficulties with Dr. Declary—told him so. "Whatever the fact may be, you are regarded by the world as one of the extreme men," the Missourian advised Clay. The moderates distrusted Cassius, he went on, and they were in control of the patronage.[10]

Rollins had given the distraught Clay an idea, and he was at the point where he would try anything. If, as Rollins said, the moderates directed the appointments, then he would become one of them. Casting about for some means of quickly winning the confidence of that wing of the party, he recognized the possibilities of the compromise agitation then going on in Washington. Clay was on record as denouncing current attempts to bring the dissatisfied segments of the Union together by political compromise. But now, in January, 1861, wracked by the dread of being neglected, he changed his tune.

Hastily, Clay set out for Washington. His purpose, he announced, was to secure a compromise which would enable border-state Unionists to resist secession. Although he urged the Republican Party to adhere to the letter of its Chicago platform, Clay said that it could, "without in the least stultifying itself, . . . grant something to the South by way of conciliation."

In the unaccustomed role of peacemaker, Clay startled his party friends, and he gained nothing by his efforts. "For the sake of our organization, for the sake of our cause, for the sake of your own future," Salmon P. Chase implored him, "give no sanction to the scheme . . . we want no compromises now and no compromises ever." Clay thus found himself in an unpopular position even among those from whom he expected reward. Lincoln himself quashed the compromise effort by declaring that he stood adamant upon the territorial issue. Republican leaders claimed that the election had turned upon the issue of unyielding resistance to slave expansion, and they would hear of no compromise with the slavery group. "I found all here fixed against any movement towards concessions," Clay plaintively reported to Lincoln.[11]

Clay found no more success as appeaser than he had as agitator; although he carefully informed Lincoln of his conservative course, his brief excursion into the paths of compromise brought him no offers. As a practical politician, Clay quickly accepted the situation. "The truth is," he confided to the President-elect, "I fear the more we concede the more will be demanded. . . ." And with that sage advice, he shelved another unsuccessful effort to obtain office. His reminders of long, unrewarded service; his militant demonstrations of loyalty; his somewhat paradoxical compromise suggestions—all had failed to win Lincoln's confidence.[12]

Inauguration Day arrived; the new administration took office; still Cassius Clay had no place in it. There were many appointive positions yet unfilled, however, and the disappointed Clay went to work to procure one of the top-ranking foreign missions. When he did, he offered another reason for seeking a prominent position: his need for money. For most of his public career, he had been plagued with debt and financial failure, and he was convinced that to escape from this burden he needed to secure a remunerative office. Shamelessly he publicized his personal difficulties to justify his claim to one of the more lucrative—and accordingly, more important —positions.

His New York friend, scholarly William H. Seward, the new Secretary of State, nominated Clay as Minister to Spain, and the Kentuckian accepted only when assured of a salary increase. He did not want the place, explaining with a sneer that Spain was an "old, effete government." But receiving the promise of an additional three thousand dollars, Clay accepted. He was glad, however, when Carl Schurz, the German immigrant revolutionary, was *persona non grata* to all European governments but that of Spain. Believing himself relieved of an inactive position, Clay resigned and accepted the ministry to Russia.[13]

Even while he was beginning to make preparations for his journey to St. Petersburg, Clay did not cease his efforts to gain a more important post as minister to England or France—or better yet, a cabinet portfolio. Once more he offered his indebtedness as justification for the request. Like a child, Cassius took his troubles to the fatherly man in the White House. "The Court of St. Petersburg is an expensive one . . . so that my family are in doubt whether it will not be necessary (seven of us!) to separate," he began. "This is the chief source of my disappointment in not having a place in the Cabinet." Then he mentioned again his protracted efforts for the party. "Now, Mr. Lincoln," the frustrated Clay pleaded, "in consideration of my life-long sacrifices, and my being again and again put back for the party's sake . . . would it be too much for me to ask of you to gratify me at least to some extent by appointing me (*in case of a vacancy*) minister to France or England?" But his supplications brought no reward. Clay was not easily rebuffed, however; amid the innumerable details of administration, Lincoln would again and again face an insistent Clay. Before Clay finally accepted his answer, Lincoln's vaunted patience would grow thin.[14]

Clay failed to receive the war portfolio, but he did not surrender his hopes for a military office. On April 15, 1861, when he arrived in Washington to receive his instructions, he found the capital in an uproar over the firing upon Fort Sumter. On the day of his arrival, the President proclaimed the existence of an insurrection and called for the state militia to sustain the national government. Cassius Clay, whose enthusiasm was notorious, gained favorable publicity by offering his services to Secretary of War Simon Cameron "either as an officer to raise a regiment, or as a private in the ranks." Cameron, surprised at Clay's ebullient patriotism, said he had never heard of a foreign minister who volunteered for the ranks. "Then," responded the doughty Cassius, "let's make a little history." [15]

After spending the day in the War Department acting as though he were its chief, Clay organized a battalion of nondescript vigilantes to guard the capital until the state troops arrived. In the confusion of going to war, rumors abounded that unnamed Confederate forces would immediately capture the city. The "Cassius M. Clay Guards" stepped into the breach and manfully prepared to defend Washington. The unit included senators, congressmen, and generals, together with an assortment of clerks and salesmen for the ranks.

As commander, Clay enlivened the atmosphere at his headquarters in Willard's Hotel with his braggadocio. With three pistols strapped to his waist, and an elegant sword hanging at his side, he talked to anyone who would listen about his Mexican War exploits and his political battles. John Hay, private secretary to President Lincoln, could hardly suppress his laughter at the droll picture Clay presented. Hay remarked that Clay ran up and down the White House steps "like an admirable vignette to 25-cents worth of yellow-covered romance." At Willard's, Hay declared, Clay spent his time talking and drinking coffee: "The grizzled captain talks politics on the raised platform and dreams of border battles and the hot noons of Monterrey." [16]

Despite the comic-opera turn Clay gave to the defense of Washington, the "Strangler Guards" command gave him opportunity to consider the advantages of military service. The Clay Battalion was more a posse than a military force, and it did more to increase the existing tension than to allay it, but it gave its captain visions of the opportunities the war would provide. With his ambitious gaze fixed upon the next election, he saw political value in a military command. His friends told him that he should go into the army, not disappear in Europe. "This war will make the next President," one of them predicted, "and I hope you will be the Major General of the Indiana troops." In addition to the Hoosier offer, Clay also received an invitation to head a New York unit. The idea of a command appealed to him, but for the present he decided to keep the political appointment. With his family he boarded a steamship in Boston, bound for duty in St. Petersburg. After many years of struggle and months of exasperating uncertainty, Cassius M. Clay was at last on his way to a public office.[17]

As his ship wallowed through the cold Atlantic, Clay had plenty of time to reflect upon his recent actions. The more he considered the turn of events, the less he relished an extended stay in distant Russia while a war raged in America. Impetuously he changed his mind. He offered to resign his mission if the President would commission him a general in the regular army. "I think my talent is military and that I will not fail the public expectation," he explained, adding that his debts influenced his decision. "I was in debt before I left home," he confessed to Lincoln, "and that is the reason why I desired a place in the *Regular* army . . . that I might know how to rely upon the means of paying off a distressing debt." [18]

The fever of ambition still raged within him, and he made arrangements for a publicity agent to "blow the Biographical Bellows" for him at home. Abroad, however, he did a good job of publicizing himself. No sooner had he arrived in England than he found much to deplore. He first complained about London's crowded hotels. After he had inquired at five of them, with nine persons in his party and four cabs loaded with baggage, he lost his temper with the hotel clerks. Benjamin Moran, secretary of the American legation in the British capital, found him at the Westminster Palace Hotel, "walking up and down the magnificent hall like a chafed lion, and looked a man to be avoided." Moran, accustomed to the stiff diplomatic costume, smiled to see a minister in a blue dress coat with gilt buttons. But Clay's anger interested him even more. "The incivility of the servants at the hotel disgusted him, and he was disposed to swear," Moran reported. After some planning, the secretary found lodging for Clay's family and left him at two o'clock in the morning, "in a humor far from pious." [19]

Not only did Clay's volcanic temper erupt at the London hotel clerks; he complained even more heatedly about the attitude of British political leaders. When he talked to Lord Palmerston, the Prime Minister, and to other influential men, he was shocked to discover prosouthern sentiments among them. Clay visited Parliament and watched in righteous wrath as the cautiously neutral members tabled a pro-Union resolution. Never one to heed the niceties of protocol, in costume or conduct, Clay undertook to enlighten the British populace. In Kentucky, whenever he was not favorably received by the men of influence in the state, he carried his case to the people through the press. Now a diplomat, accredited by the United States government, Clay employed the same tactic. Going over the heads of the British leaders, and blithely ignoring Minister Charles Francis Adams, Clay published a vigorous statement in the London *Times*. The war in America, he informed the British, was rebellion; the term "secession" concealed treason. Despite Lincoln's insistence that the purpose of the war was to restore the Union, Clay assured his readers that it was an antislavery crusade, and that England's honor required that she support the Union cause. With typical ferocity, Clay threatened the British as though they were Kentucky slaveholders. "England, then, is our *natural ally*," he concluded. "Will she ignore our aspirations? *If she is wise, she will not*." Cassius Clay, who, as a rough-and-tumble fighter, had become a master of the sharp retort and the aggressive insinuation, brought the same methods to the foreign service. As he had gained fame as a bowie knife emancipationist, now he attained a reputation as a bowie knife diplomat.[20]

Unfortunately, however, Clay was not dealing with provincial Kentuckians who respected the law of the knife, nor did he enjoy the liberty of a private citizen. Instead of winning friends by his hint of violence, Clay merely brought ridicule upon himself and upon the Union cause. Americans in London reported that his spirited outburst had amazed the English, who regarded his conduct as unbecoming to a diplomat. But Clay's mail swelled with complimentary notes from antislavery Republicans, loud in their praise.[21]

Despite the criticism, Clay was quite pleased with his introduction to international affairs. From London the Clay entourage moved on to Paris. In the land of Napoleon III, Cassius was again shocked to discover that a firm Union policy was lacking and that Confederate agents were making headway with the French government. Again he wasted no time with the niceties of diplomatic procedure. On the morning of May 29, the "American citizens in Paris favorable to the Union" breakfasted together at the Hotel du Louvre. A host of itinerant Republican ministers gathered to hear Clay's fiery speech. This time the Minister to France, William L. Dayton attended the festivities. Also present at the speakers' table were Republican workhorse Anson Burlingame, who had been rejected at Vienna as too radical for the Austrians and was now on his way to the Chinese legation; Jacob S. Haldeman, of Pennsylvania, Cameron's friend who had received the appointment to Stockholm; and John C. Fremont, 1856 Republican candidate for the Presidency, who was in France seeking funds for his rich California mining property. Since Dayton had been the vice-presidential candidate on the same ticket with Fremont, both Republican standard-bearers of the previous election attended the Paris breakfast.

Before that distinguished group of party leaders and appointees, Clay reiterated his warning to England, and reminded France by pugnacious threat that her interests lay with the Union. He appealed to traditional French animosity toward the English, who might try to "mingle the red crosses of the Union Jack with the piratical black flag of the Confederate States of America."

Such outbursts in London and Paris irritated the Adams family in the London legation. "Those noisy jackasses Clay and Burlingame," Henry Adams complained to his brother, "have done more harm here than their weak heads were worth a thousand times over." Clay continued to gather publicity —some of it unfavorable—by his bold, undiplomatic course of American spread-eagleism.[22]

But once in Russia, Clay learned that the manifestoes which made other ministers' tasks more difficult proved to be assets to his own mission. His instructions were to win the support and the sympathy of the Russian government, and his "athletic Western argument" (as the eccentric railroad promoter, George Francis Train, called it) helped him to accomplish that end. Because the English were unpopular in St. Petersburg, the unrestrained declarations he had made in London and Paris simplified his official task. Indeed, as Minister to Russia, Cassius Clay was an indubitable success. He carried out his instructions more completely than did any of his diplomatic colleagues in western Europe. The dignified Prince Gortchacov, Prime Minister of the Czar's government, assured Minister Clay that Russia's official sympathy was with the Union.[23]

Despite his successes, Cassius was not satisfied to remain in exile so far from the scene of action. Night after night in the summer of 1861 he sat at his desk penning ponderous letters to the Secretary of State by the light of the midnight sun. He gave Seward, or Cameron, or Lincoln, gratuitous advice upon the conduct of the war as well as upon a thousand other irrelevant matters. Meanwhile, he won friends in the Russian capital by prodigious entertainment. His parties were the most brilliant spectacles he could arrange, with the best wines in abundance, exotic foods, and lavish entertainments. "I was determined to please," was his solution to the problems of diplomacy. From the generous hospitality of Kentucky aristocracy it was but a short step to the polite society of Russian nobility, and Cassius and Mary Jane Clay adjusted easily.[24]

Clay felt at home among the Russian noblemen, who were as yet serf-owners and, he said, were little different from American slave-owners. He acted as though he were back in the Kentucky Bluegrass, and he never went out unarmed. He had brought bowie knives of every variety with him; the ferocious blade remained his favorite weapon. He had a pearl-handled knife with an eighteen-inch blade for formal attire, but for everyday wear he had an assortment of bone-handled knives. It was not long before his pugnacity had brought him challenges to duels from rapier-bearing Russian noblemen. As he thus had the right to choose the weapons for the fight, he always selected the bowie knife. This choice upset the suave swordsmen, and they plotted to trick Cassius into making the challenge so they would have the choice of weapons. Then, they promised themselves, Mr. Clay had better beware.

Having made their plans, two Russians found the American Minister in a café, eating dinner. One of them slapped Clay on the cheek with his glove and then stepped back to receive the Minister's card and a challenge for a duel. But Cassius Clay was never one to stand upon a precedent; he owed his life to his ready defense. He jumped up from his chair, doubled up his huge right fist, and with the full force of his strength hit the meddlesome Russian in the nose. Overwhelmed by the unexpected blow, the trouble-seeker went down, carrying a table with him amid a storm of falling china. Clay looked around belligerently for further opposition, and seeing none, parted the tails of his frock coat and sat down to resume his interrupted meal.

He was gathering a reputation for himself in St. Petersburg for his quick pugnacity, which aroused the admiration of his Russian hosts. Stories of his bloody fights circulated over Europe as well as back home in Washington. Such affairs gave substance to Bayard Taylor's derogatory opinions of Clay. Taylor, secretary of legation after Clay left Russia, was jealous of anyone who might prevent him from succeeding to the position of minister, and he wrote complaints about Clay to influential Americans who might carry weight with Lincoln. "Between ourselves," he told Horace Greeley, speaking of Clay, "he is much better suited to the meridian of Kentucky than of St. Petersburg." [25]

Minister Clay did, however, win the approval of his Russian acquaintances, and he was adept, too, at the requirements of court etiquette. On Sunday, July 14, 1861, the new American Minister was presented to the Czar. Clay dressed himself in his military regalia, with a sword knocking at his ankles. He took with him his two secretaries, Green Clay, his nephew, and William Cassius Goodloe, son of Clay's friend and business associate, David S. Goodloe. After a short train ride the party arrived in Peterhoff, where three imperial carriages met them and took them to the palace. There they passed through a guard of soldiers in colorful uniform, and they participated in a review of the troops before the Czar arrived.

To meet Alexander II, the Americans were escorted through several large rooms; then they met a man who had "enough feathers in his hat to make an ostrich," Goodloe remembered later. Guided by the "feathered individual," Clay went into the presence of the Czar, where the Prime Minister, Prince Gortchacov, introduced him to Alexander. Cassius presented his credentials and made a "set speech," to which the Emperor responded by pledging the continued friendship of his government.

Alexander II, Czar of All the Russias, cut a handsome figure. He was dressed in military uniform, Goodloe recalled: "Blue coat, military buttons, blue trousers, small gold stripes, calf skin boots." The young secretary described the Emperor as "stoutly built, and of an exquisite figure. Very handsome, rather a round face, eyes a beautiful light blue, mustache, hair shingled, and of a dark auburn color. Speaks 'American', voice pleasant, and looks and walks and is, every inch a King."

After the meeting with Alexander, the Americans were led in to lunch. Goodloe was amazed at the splendor of the meal: "Soup, chicken, chops, strawberries, oranges, peaches, with six brands of wine." When the lunch was over, the guide took them to the old palaces of Peter the Great, and they spent the afternoon going through the museum-like buildings. Clay, who enjoyed the royal treatment he received, became a staunch admirer of Alexander II and of his government. But his approval of the Emperor led him into one grave mistake: he modelled his own parties after the lavish scale of imperial entertainment, and he soon discovered that he could not maintain that standard.[26]

His expenses confirmed Clay's predictions that his salary was inadequate. As he saw himself fading from the political scene back home, once more he took up his pen to request a more prominent position, and he offered the excuse of his financial difficulties. "It is not possible on trial for us to live here on 12,000$," he complained to the President. "If my salary is not raised, I shall be forced to return home." To emphasize his words, in the fall of 1861, he sent his family home (Mary Jane did not like the damp, frigid climate anyway), and made plans for a more penurious existence. Then, just as he had completed his arrangements and purchased house furnishings, he received word that the President had commissioned him a major general of volunteers.[27]

He had repeatedly requested a military commission, but when he received one, he was dissatisfied to learn that it was a temporary, not a permanent, appointment. He feared that the war might soon be over, and he would be forced to retire to private life. He kept up his demands for a cabinet post. "I don't see where I am to take command even if I were now at home," he grumbled to Lincoln. The "foremost and most desirable" places were already filled, he pointed out, and he had a dread of being assigned to a "temporary and inert" command.[28]

Simon Cameron, who had been shifted from the War Department, was to be Clay's successor. When Cameron arrived in St. Petersburg, Clay was pleased to discover that the Pennsylvanian intended to remain in Russia only long enough for the storm of War Department mismanagement to blow over. That intelligence gave Clay a chance to plan a retreat should he be shunted to an inconspicuous military outpost. "As you have now too many generals in the field," he pointed out to the President, "I ask that you will return me to this court after Mr. Cameron's leaving." Clay did not want to leave his civilian office without some means of returning to it if need be. Reluctantly, the President agreed to his request and committed himself to return Clay to Russia, if he still wanted the post when Cameron resigned it.[29]

Clay now had a double assurance against disappearing from the public scene. If he received an active command, he would receive much publicity which would be useful to him; if not, he had the President's promise to send him back to St. Petersburg. Although it was a distant post, at least he would not be out of the political realm altogether.

With a major general's commission and Lincoln's promise in his pocket, Cassius Clay returned home from his ministry to Russia. The next few months were extremely important. As he journeyed across Europe, he was preparing for the intraparty political struggle which lay ahead.

CHAPTER XIII

CLAY'S STAR FADES

WHEN Cassius M. Clay returned from his mission to St. Petersburg he opened an active campaign for leadership in the Republican Party. That organization, in 1860 a hodgepodge of political malcontents, presented no solid front two years later. One wing of the party, composed largely of ex-Whigs, was conservative and looked to the party for economic legislation friendly to the business community. Another group, which had entered the Republican fold through the abolition door, wanted to use the party victory to demolish the "relic of barbarism," chattel slavery. Although on economic matters they were as conservative as their ex-Whig allies, they were known as "Radicals" because of their uncompromising war upon slavery. The antislavery crusade had served the purposes of the more practical-minded industrialists, and the two forces had fused to win the election, but between them there remained friction. Cassius Clay, as a man who favored emancipation of slaves because of hardheaded economic considerations, was a bridge between the divergent wings of the party. His choleric temperament, however, gave him a reputation as a fire-eating extremist. Although not as radical as some of the abolition Republicans, still he was making demands that the war be made an antislavery instrument. To rise in the party ranks, now dominated by a shrewd President, he had to win the support of the disgruntled antislavery Republicans who criticized Lincoln. Clay was a problem to the President, but Lincoln foresaw the threat and moved to meet it. His handling of the explosive Clay was an example of his finesse as a politician.

A week after he arrived from Europe, Clay voiced his disapproval of the administration's management of the war. On August 12, 1862, he opened his campaign for leadership among the Radicals with a public excoriation of Lincoln for protecting slave property. That was a "wishy-washy, milk-and-cider" war policy, he said. "You are going to conquer the South by taking the sword in one hand and shackles in the other," he gibed to his Washington audience. "You are going to conquer the South with one portion of your force, while the other is detailed to guard rebel property." Such a policy, Cassius said, was an anomaly. "Better recognize the Southern confederacy at once, and stop this effusion of blood," he advised, "than to continue in this present ruinous policy."

He held a commission as major general of volunteers, Clay announced, but he would participate in no military action which did not anticipate the destruction of slavery. In order to draw Radical support he took an extreme position. "I shall strike only for liberty, and will never draw the sword for the protection of rebels' slaves," he defiantly declared. He expected no success in the army, he fumed, with a hostile government at his back; if he were to carry out his views in the field, the administration would "shelve" him, as it had removed John C. Fremont from command a few months earlier.

Although Clay ardently criticized Lincoln's war policies, he avoided an open break with the President. He held the President to his promise to return him to Russia. "Although you may not always represent my special views," he told the President privately, "you have always my *confidence and support* to carry out your own—for you are the Chief of the Nation—not I." Even from the beginning, Clay's revolt was a weak one.[1]

Clay quietly acknowledged the President's power, but he won the compliments of Radical politicians by his sarcastic criticism of the administration and its war effort. Hotheaded Kansas Senator Samuel C. Pomeroy, who would later win a dubious notoriety by endorsing Salmon P. Chase as the Republican nominee in 1864, wildly praised Clay for his outspoken comments. "This vacillating policy—or want of a fixed policy—is demoralizing the nation!" Pomeroy declared. And in Boston, the abolitionist and orator Wendell Phillips agreed with Pomeroy's enthusiasm. "It seems to me," Phillips assured Clay, "*you* have the power to *hasten* the adoption of the needed policy so much as to save thousands of lives, millions of dollars, and untold dangers to Republicanism. . . ."[2]

Despite such ebullient commendation from the Radical fringe of the party, Clay met only apathy from the public. His campaign, begun with much ardor, was already failing. The press was unfriendly, and poked fun at him for trying to assume the President's responsibilities: the *New York Times* expressed ironic sympathy that the war did not suit Clay's tastes, and hoped that he would find a place where his "very great ability" would prove useful. Clay's demand that the war be used to destroy slavery drew upon him the opprobrium of citizens who considered the preservation of the Union sufficient cause for which to fight.

But Clay saw himself as representing a great segment of public opinion. "If I were to carry out in the field my political views I might rub across the President," he confided to the new Secretary of War, Edwin M. Stanton. "Should I not be allowed to do so, it might array against him a larger party who would otherwise give him aid and comfort." Consequently, he planned to return to Russia, "at some sacrifice," he informed his chief, Stanton. His decision indicated that his political aspirations were not going well.[3]

Though Clay exaggerated Radical strength, Lincoln recognized that it was a minority and that most northern citizens favored a more lenient policy. But the President was too astute a politician to make martyrs of his Radical critics by persecuting them. Instead of dignifying opposition forces by replying with an indignant rejoinder, he was accustomed to draw the fangs out of criticism and to continue to utilize the critic. Clay's quixotic rebellion Lincoln deflected in the same manner: he sent the Kentuckian upon a harmless mission which won him over to the President's camp.

For months the President had wrestled with the dilemma posed by the demands of the Radical abolitionists in his party, who wanted him to adopt a firm antislavery policy, on the one hand; and the threats of the sensitive border-state slaveholders, who might secede if the war became an emancipation crusade. In the summer of 1862, Lincoln decided to take some action to appease the abolitionists, and he sought as harmless a gesture as possible: to proclaim emancipation for all slaves in rebel states, but to protect the slaves of loyal masters behind federal lines. This solution would make emancipation an official war objective, without frightening the border-state slave-owners into secession. In July Lincoln brought up the matter in his cabinet meeting, but when some members objected to it, he agreed to let the matter rest. But when dissident Radical governors planned a protest meeting for September 24, at Altoona, Pennsylvania, he saw that he would have to act soon. Never one to await events, Lincoln planned to take the initiative in his political war with the Radicals.[4]

At the moment, the President also had a critical Cassius Clay upon his hands. Lincoln surmised privately that Clay was an egotist, proud of his reputation, but petty and mean if crossed. Clay possessed a "great deal of conceit and very little sense," Lincoln confided to his old friend, Orville H. Browning. Wisely, the President played his hand to forestall criticism from the governors' conference, and at the same time to nullify Clay's incipient revolt. He played upon Clay's self-esteem. "I have been thinking of what you said to me," Lincoln began, as though emancipation by Presidential proclamation were Clay's idea. Then he explained his proposition. "But I fear if such proclamation of emancipation was made Kentucky would go against us, and we have now as much as we can carry."

Cassius promptly arose to the bait. He was pleased to consider himself an adviser to the President. "You are mistaken," he responded. ". . . Those who intend to stand by slavery have already joined the rebel army; and those who remain will stand by the Union at all events." Clay, who had not been in Kentucky for fifteen months, based his hasty assertion upon hearsay evidence, and Lincoln knew it. But instead of questioning him, the President dispatched him upon an errand. "The Kentucky Legislature is now in session," he reminded Clay. "Go down, and see how they stand, and report to me." Lincoln, who had already decided to take action, reasoned that with a hand in the forthcoming pronouncement, Clay would be predisposed to accept it. Moreover, the expedition would keep him busy. He intended to go to Kentucky anyway, and now he would go as the President's representative.[5]

When Cassius reached his home state he was temporarily diverted into military service. In Cincinnati he learned that Confederate troops under Kirby Smith had invaded Kentucky through the Cumberland Gap and were headed toward the Ohio River. Offering his services to General Lew Wallace, Clay received command of volunteer infantry and artillery troops, with instructions to repel the invaders. He was familiar with the terrain along the Kentucky River and he planned to utilize its steep banks as part of his defense line. Slowly he marched his raw troops toward the river. Before he completed his dispositions, however, he was rudely relieved of the command by General William Nelson, who had been an officer in the regular navy. The new commander impatiently pushed his men across the Kentucky, and on the Richmond plain in central Madison County they met Smith's superior forces. There the federal troops, exhausted by forced marches, sustained one of their worst defeats of the war. With no reinforcements, Nelson's luckless command was unable to hold the line of the Kentucky River.[6]

While the battle raged, Cassius Clay was in Frankfort, his brief military service in the Civil War ended. When he was relieved of his command he returned to his original mission. He addressed the Kentucky legislature in the State House and released Lincoln's trial balloon, but he explained it as though it were his own idea. The federal government, Clay began, had no power to war upon slavery, or upon any other form of property, but rebels forfeited their rights to life, liberty, and property. "In the loyal slave States," he continued, "I would not injure the loyal slave-owner . . . but in the rebel States, I would proclaim liberty to the slaves of disloyal masters. . . ." In brief, he summarized, he would "recall the four millions of black allies whom, in a false magnanimity, we have loaned to the enemy." Should the proposed emancipation program injure any loyal slaveholders, he would recommend, Clay said, that they receive compensation from the national treasury. As a messenger from the President, Clay revealed an unaccustomed moderation. If rebels would lay down their arms and return to their allegiance, he would propose the repeal of all confiscation acts, he said, and urge a general amnesty. Lincoln's little trick had worked. As his spokesman instead of his critic, Clay had removed himself from among the opponents of the administration. Moreover, the subsequent proclamation of emancipation would come as no surprise to the Kentucky lawmakers, who welcomed Clay's conciliatory remarks.[7]

Clay lost no time in reporting his success to the President. As Kirby Smith's Confederate Army was approaching Frankfort, the departing Clay had plenty of company along the road to Cincinnati. From there, Cassius went on to Washington to tell Lincoln that Kentucky citizens would accept a proclamation freeing the slaves of rebels. The President said nothing to Clay about his plans, but a few weeks later, on September 22, he issued the preliminary Emancipation Proclamation. Clay, who never saw the connection between the meeting of disgruntled governors at Altoona and the "humanitarian" measure, assumed a share in the President's action. Indeed, Lincoln's political prestidigitation fooled Clay completely. Clay regarded the proclamation as "immortal," and he later exulted that his "good star stood high in the heavens; and whilst my enemies sought by unworthy means my ruin, I seemed by Providence to have been called for the culminating act of my life's aspirations." But as his ambitions were political rather than humanitarian, the appearance of the proclamation signified that Clay's "good star" had faded.[8]

Deftly the President had cut the ground from under the Radical governors, and he had won Clay to his side. No longer a critic, Cassius now became a supporter of the administration's war aims. On the night of September 24, a procession of jubilant citizens serenaded antislavery politicians in Washington. From the White House, the parade moved to the residence of Salmon P. Chase, and then to Clay's hotel. Honored by the show of tribute, Cassius endorsed Lincoln's course of action. Six weeks earlier, returning from Russia with ambitious fire in his eye, he had just as enthusiastically denounced it. His rebellion had ended. Later, Lincoln carried his little joke a step further when he sent the admiring Clay a portrait of himself holding the Emancipation Proclamation. By superior political ability Lincoln had won control of his party.[9]

Clay recognized that his chance for leadership among the Radicals had disappeared, and he resigned his military commission to clear the way for his return to Russia. He had given up all hope of getting a prominent command, and the Emancipation Proclamation had destroyed his differences with the administration. Since Cameron had not yet resigned, however, Clay requested permission to embark upon a speaking tour on behalf of the 1862 Republican candidates. The President, having converted Clay to his side, willingly ordained him a prophet of Republicanism.[10]

But though Lincoln blessed the undertaking, Clay aroused little attention, and his best efforts had no effect upon the fall elections. He spent most of his time in New York, in an effort to arouse opposition to the Democratic candidate for governor, Horatio Seymour, and to his slogan, "The Union as it was." Clay's speeches were pitifully hackneyed, and only partisan audiences appreciated them. He indulged in prolix recitals of his long experience as an agitator, and he boasted of the time when he "stood entirely alone in the defense of these same principles, threatened again and again. . . ." No longer did Clay attempt to justify his action on economic grounds, as he had earlier; now he declared that he had opposed slavery merely because it was a moral evil. "I came to the conclusion that I was right," he confessed, "and based upon that eternal basis, I have, from that day to this, contended for those principles which I trust I may yet live to see carried out. . . ." Garrulous autobiography became Clay's main theme.

In addition to proclaiming his loyalty to freedom, Clay denounced the Democrats for urging a return to the Union as it was. "What was it to me?" he demanded. "The scars upon my body testify that it made me a slave." He had lost none of his vindictiveness: "I believe that hanging such men as Seymour and Van Buren would save many a good Democratic life," he declared in New York, where those men were candidates. But with all his fervor, Clay offered no program beyond emancipation. Even his oratory had lost its fire. Despite his efforts for the New York Republican candidates, Seymour won the election. War-weariness and a general distaste for the Emancipation Proclamation were responsible for the Democratic victory. The election demonstrated that a greater political wizard than Cassius Clay would be necessary to win in 1864.[11]

Clay was no longer the popular champion he had been earlier. His vigorous declarations did not attract the attention they once had; he was out of touch with the ground swell of sentiment which accompanied the war. Clay's inability to change was his greatest weakness. In 1845 he had been a national hero in the cause of civil liberty, but he never grew beyond that position. For twenty years, because of his extreme Whig views, he had declared himself an emancipationist, and in 1863 he beat the same drum, worn thin with use. His hunger for prominence made him a pathetic figure; in the rude helter-skelter of politics, he was a has-been. At this time a people wrestling with itself in fratricidal strife, struggling to meet the challenge of forming a "new nation" based upon a changed coalition of forces, demanded more of its leaders than trite repetition of out-of-date antislavery contumely. Though he continued to draw respectable audiences, he told them nothing he had not said many times before. The Emancipation Proclamation had left Clay no further program to offer.

Lincoln's document proved an obstacle to Clay's ambition, but the key to his failure lay in the limitations of his character. The antislavery phase of his career had become a monomania, and he closed his eyes to other issues. After emancipation was effected, the remainder of Clay's public career was the sad tale of a man whose single purpose in life had been accomplished. Living in the past, Clay was dependent upon his record for recognition; he sought attention by claiming that victory over slavery was his success. But he could go no further. His inability to grow beyond his obsession with fighting slavery doomed Cassius Clay to failure as a politician.

After his ineffectual participation in the New York campaign, Clay surrendered his aspirations for the Presidency and accepted the only office open to him. When Cameron returned from a tour of duty in St. Petersburg, Clay reminded Lincoln of his promise to send him there again. Though the President regretted giving his word, he lived up to it and recommended Clay for the post. Clay's earlier undiplomatic antics had aroused serious opposition to his reappointment; some senators charged that he was unpopular even among the Russians. The *New York Times*, no friend of Clay's, had sarcastic comment about the Russian mission: "We hope that the Emperor will not get the impression that we confound his dominions with Siberia, and regard them as penal colonies for the banishment of uneasy politicians. . . ." Of Clay, the *Times* editor declared that "nature had indicated in the clearest manner, that she did not mean him to be a diplomatist, and his education has not been of a kind to lessen the forces of her prohibition."

With such barbed remarks in the press, and with determined opposition to him in the Senate, Clay feared that he would not be reappointed. But Lincoln let it be known that he personally desired Clay's confirmation, and the Senate complied. Though it was clear that the President had influenced the vote, Clay boasted that it was the "greatest triumph of my life." For a man known to have had presidential aspirations, that confession was a tacit admission that he no longer figured as a contender for high office.[12]

After his confirmation by the Senate, Clay prepared to return to Russia. His wife and family were to remain in

Kentucky, and he had a lonely time ahead. With his ambitions for political prominence dashed, his only reason for accepting the mission was to pay off his debts and make plans for his retirement. Well aware that his salary would be slim, Clay began to hint at "deals" in which he might earn commissions and fees to supplement his official remuneration. Cautiously he broached his plan to Salmon P. Chase. "You know how hard I have been down in debt," he reminded the Secretary of the Treasury. "Is there anything, in which you can serve me . . . in your financial arrangements at home or abroad?" In addition to approaching Chase, Clay allowed the use of his name to endorse the sale of stock in the Union Pacific Railroad. Clay may have been out of touch with the political realities of the day, but he was well aware of prevalent economic factors. Leaving behind contacts which might develop into large extraministerial business, Cassius M. Clay returned to St. Petersburg.[13]

CHAPTER XIV

SWASHBUCKLING DIPLOMACY

NOW fully aware that he had no chance of obtaining the 1864 presidential nomination, Cassius M. Clay returned to the Russian mission, all that remained of his lofty ambitions. But he consoled himself that his old enemy, slavery, was on the way out. "To me the final triumph of my principles was of more worth than elevation to office," he explained later. With that salve for his pride, Clay left the American scene. For more than six years, through the administrations of Abraham Lincoln and Andrew Johnson, and into Ulysses S. Grant's regime, he remained in Russia.

But his second term was less satisfying to him than the first had been. Lonely and frustrated, Clay was troubled with many blunders of indiscretion which marred his social standing. Violent fits of jealousy, bitter tirades against his enemies, and the dark shadow of scandal beclouded his stay in St. Petersburg. Clay's swashbuckling nature asserted itself while he was chief of the legation, and his diplomatic career illustrated Russian tolerance of an often tactless, imprudent, blustering minister.[1]

Clay's eccentricities did not immediately appear, and at the beginning of his second term as Minister Plenipotentiary and Envoy Extraordinary to the Court of the Czars he carefully attended his business. In April, 1863, he moved into a house on Golanaia Street, "near the Peter monument," he wrote to Mary Jane, who, despite his entreaties, had remained in Kentucky. In that house Clay entertained Russian society, and from it he journeyed in his carriage to the legation and on sight-seeing trips. The Russian capital was built upon marshy, uninviting land near the Baltic Sea, but it had been developed until, when Clay was there, it was the imposing capital of a sprawling empire. Czar Alexander II was in the process of liberating the serfs, and Clay, who considered himself an emancipationist, felt at home as representative to the Czar's court.[2]

The European situation also eased his diplomatic task in Russia. His primary mission was to maintain friendly relations with the Russian government, and for that purpose the Polish insurrection then going on served handily. In November, 1860, Polish nationalists revolted against the Russian force occupying their country, but for some time the affair aroused little attention. When early efforts to put down the rebellion failed, however, and the Russians resorted to terroristic tactics, there was an immediate response from the anti-Russian bloc in Europe. For a time there was fear that England would intervene on behalf of the embattled Poles. As England's dependence upon raw cotton made many of her citizens prosouthern in the American crisis, both the United States and Russia had reason to mistrust the British. Their mutual antipathy drove the two governments into each other's arms.

In the diplomatic conflict, Seward's policy of playing off the Russians against the British became clear. The rulers of Great Britain, France, and Austria requested the United States to join them in protesting Russia's handling of the Polish affair. In rejecting the offer, Seward used the opportunity to push Russia and England further apart, and he composed a note filled with warm avowals of Russian-American amity. Clay had Seward's note published in Russia, and, as the Secretary of State had intended, it won Russian approval. Like a good diplomat, Clay found excuses to prefer in the dispute that side which benefited his country. "Our interests," he assured Seward, "are on the side of Russia against reactionary, Catholic Poland." [3] Because the Czar pursued a program of emancipation, Clay described the Russian government as "tolerant in religion, and progressive in civil administration." Emancipation, whether by presidential proclamation or by imperial ukase, was enough to convince him of a government's merit.[4]

Despite the public avowals of Russian-American friendship, there was still a threat of English and French intervention in Poland. "All Russia is aroused up to the defence of her territory," Clay reported, "and she will make an 1812 affair of it before they get through if France and England force a fight upon them." In the crisis the Russians prepared for war. Realizing that their fleet would be bottled up if hostilities broke out, they sent it to the comparative safety of American ports. ostensibly upon a good-will tour. As no other European power made so tangible a show of friendship during the Civil War, Americans readily interpreted the visit as evidence of Russian sympathy for the Union cause. Thus, the existence of two widely separated rebellions, which had only one factor in common—the fear of British intervention—made Clay's diplomatic task simpler. Through no unusual action of his own, but because a common enemy confronted the United States and Russia, his ministry to the Court of the Czars was a success from the beginning. His bold twisting of the British lion's tail endeared Clay to Anglophobe Russians, and as a representative of a country sympathetic to Russia in the Polish rebellion, he could expect to be well received.[5]

Clay's diplomatic post was not a difficult one, but in his own official family he had trouble with his subordinates. Though it sprang largely from his own jealous attempts to safeguard his ministerial prerogatives, he blamed Seward for sending "spies" into the legation to "calumniate" him. He repeatedly charged that Seward was his bitter enemy because the Kentucky and Virginia delegations at Chicago in 1860 had supported Chase and then Lincoln. "Ever since I refused to take part for Seward as President in 1860, he has been my enemy," he reported to Massachusetts Senator Henry Wilson. Because the sharp-featured Seward was Clay's superior, Clay blamed him for all his woes, including his legation problems.[6]

In 1863, before he left New York, Clay had asked for the appointment of Henry Bergh, philanthropist and author, who was later to become founder and first president of the American Society for Prevention of Cruelty to Animals, as secretary of the legation. But when Bergh called upon Seward about the appointment, the Secretary of State received him brusquely. Once more Clay took his complaint to the man in the White House. "You see all this is merely a pretext to insult me, by insulting my friend," Clay argued. "He is insulted: and I am to be put in the position of having a Seward Spy in my house. . . . If a Sewardite is forced upon me," he went on, "I shall regard it as an unfriendly act on your part." Lincoln refused to enter the intramural controversy, but Seward did allow Clay a free hand in the choice of his assistants.

Still Clay was not satisfied. Bergh got the secretarial post, but he did not keep it long. Because of Clay's apprehensions that one of his underlings would become more popular with the Russians than he, and take away his position, he mistreated them all. Clay's jealousy erupted, and Bergh soon returned to the United States.[7]

The next secretary, who held the position until 1869, was young Jeremiah Curtin, a native of Wisconsin and a Harvard graduate. Although to the uncomplimentary Benjamin Moran in London he was a "silly looking fellow with his hair parted in the middle," Curtin possessed a native talent for language and had taught himself Russian. When the Czar's fleet docked in New York, he made friends among the crews in order to practice his pronunciation, and they invited him to Russia. Before the end of 1864 he was in St. Petersburg, and he applied for a job at the American legation.[8]

Clay hired him to fill the vacant position as secretary of legation, and was immediately pleased with his find. When he presented Curtin to the Czar, Alexander addressed the young man in the customary French, but the secretary amazed him by responding in Russian. Such an unusual occurrence attracted the Emperor's attention, and Clay later exulted to Curtin, "You have made a great hit, a great hit!" For several years Clay and Curtin worked together amicably, the newcomer's knowledge of Russian proving useful to the Minister.[9]

Clay made use also of Curtin's understanding of the Russian people. Curtin had not been in St. Petersburg long before Clay brought him a problem precipitated by the Minister's own undiplomatic blustering. In pressing the claims of the Russian-American Telegraph Company Clay had been a bit too strenuous, and he feared that he had insulted the Russians. A scheme to connect San Francisco and St. Petersburg with a telegraph line by way of the Aleutian Islands and Siberia had been originated by a United States commercial consul in Eastern Asia, Perry McDonough Collins. Collins had enlisted the aid of Hiram Sibley, president of the Western Union Telegraph Company, and together the two had organized a new company. In 1862, when Collins went to the Russian capital to secure a charter for the work, he ran into trouble. The Russians were willing to charter the company to operate in their country, but they demanded an unusual share of the profits—or so Collins declared. The following spring, when Clay arrived, he became Collins' representative and heatedly demanded that the Americans be given more control over the enterprise. So excited did he become that when he talked to Prime Minister Gortchacov he was stubborn to the point of rudeness, and he came away worried for fear the Russians might demand his recall. He took his troubles to Curtin. "You know the Russians much better than I do," he confessed. "What can be done to conciliate Gortchacov?"

After much deliberation, Curtin advised Clay to make an agreement with the Prime Minister to let Seward settle the matter. Curtin recommended that Clay tell Gortchacov, "Personally, we are good friends, but in this Alaskan affair we are of different opinions." Curtin's solution worked. Gortchacov accepted the compromise, the Secretary of State made satisfactory arrangements, and work began on the communication line.[10]

Clay's interest in the company was not entirely that of a disinterested public servant. While Collins and Sibley were in St. Petersburg, Clay invited them to dinner every Sunday and introduced them to polite society. His efforts were not unrewarded. The promoters gave Clay thirty thousand dollars worth of paid-up stock in the company, intrusted to him several hundred thousand dollars worth of ordinary stock to distribute as grease for the wheels of Russian influence, and gave him additional stock to sell on commission. Clay first sold his own paid-up shares and applied the proceeds to his debts in the United States. Later, the company failed because of competition with the Atlantic cable, which was completed first, and several Russians accused Clay of cheating them when he would not refund their money. Though the enterprise created some ill will among unfortunate Russian investors, Clay made money from it.[11]

Clay profited from other financial arrangements in Russia; not only did he receive "presents" for his influence as minister, but he also used his money to speculate in bonds on the United States market. After several years he paid off the last of his distressing debts, and his main objective in accepting the ministry was achieved.[12]

While Clay made money in Russia, he steadfastly resisted an attempt to force the Russian government to pay an American claim. He wanted to maintain friendly relations with the official Russians, and he considered that the claim was a swindle. The Perkins Claim, as it was called, had arisen during the Crimean War, when the Russians were hard-pressed to maintain a minor battle-front. Russian military leaders had procured supplies in the United States, and a Boston merchant, Benjamin Perkins, had made a verbal agreement with the purchasing agent to sell powder and small-arms. But before he had delivered any of the goods the war ended, and the buyers no longer required them. The Russians peremptorily cancelled the agreement. Perkins, arguing that the verbal agreement was a legal contract, demanded that the Russians reimburse him for his loss, about fifty thousand dollars. He brought suit in a New York court, where the matter was dismissed with a settlement of two hundred dollars.

There the claim rested until 1860, when a certain Joseph B. Stewart inherited it, and organized a company, capitalized at eight hundred thousand dollars, to press it. The Perkins Claim was no longer a small affair. During his first term in Russia, Clay had received it but had indignantly refused to press it. (For that independence, he later declared, Seward used his offers to accept a military commission as the excuse to recall him.) When he returned to the Russian capital, the claim came up again.

This time, fearing a censure from Washington, Clay took the documents to Gortchacov. The Prime Minister became very angry when he read the claim, turned red, and declared, "I will go to war before I will pay a single kopeck!" Triumphantly, Clay returned the papers to Washington, blaming the graduate of the "corrupt Albany-school of politics" for the rebuff.[13]

Such difficulties, usually connected with the duties of a minister, were minor worries for Clay, compared to his losing caste with the Empress and thus jeopardizing his social standing among his hosts. From the beginning of his sojourn in the Russian capital he had been on unusually friendly terms with the social leaders of St. Petersburg, and especially with the royal family. In the winter of 1863, he boasted to Mary Jane of his success with the Czarina, and he said that the grand master of court ceremonies, Count Orloff Davidoff, had complained of it to him. "I have a grievance against you," Davidoff told Clay. "Indeed," responded the American, somewhat embarrassed. "Yes, you have superceded me in the Empress' esteem . . . you are quite a favorite with her. . . ."[14]

But an act of misunderstood gallantry put an end to Clay's favored position with the Empress. On a summer visit to the Emperor's vacation estate he went for a drive in a borrowed royal carriage. Meeting one of the princesses just as a sudden storm blew up, Clay let her have the carriage while he awaited its return—as, he protested, any gentleman would have done. But gossipers saw the girl in the carriage, and after that the Empress would not speak to the American. One of the princes heard his story, and accepted his version of what happened, but he smiled as if to say, "It's all over with you!"[15]

Clay's social life had suffered a distressing blow, but he continued his efforts to win friends by lavish entertainment, and he did not restrict himself to male company. "I have always enjoyed the society of intellectual women more than that of men," he confessed later, and his engagement calendar in St. Petersburg confirmed his preference. Along with official commands to assist the Emperor at military reviews and maneuvers, Clay's mail included small perfumed notes in dainty feminine French. In his early fifties, robust and healthy, Clay struck a handsome pose in his resplendent major general's uniform, with a jewel-encrusted sword at his side.

In his colorful costume, with his whitening hair camouflaged by dye, Clay moved among Russian society like an overgrown boy. When he saw a lovely woman—and he was easily convinced that Russian women were the world's most fair—he promptly said so. Such unconstrained compliments won the admiration of the ladies. And Clay invited them to his home, where he won their praise with his heavily liquored version of Kentucky punch. He won a widespread reputation as a Don Juan. A friend in London, who had been on his legation staff, teased him about his social aspirations. "How are you, my dear General?" he inquired. "Still making your courtesies to the bonny lasses?" But Bayard Taylor, ex-secretary of the legation, in a bad humor because he had not been promoted to minister when Cameron left St. Petersburg, was bitter about Clay's romancing. "A man (*entre nous*) who made the legation a laughing stock, whose incredible vanity and astonishing blunders are still the talk of St. Petersburg . . . will probably be allowed to come back to his ballet girls (his reason for coming) by our softhearted Abraham Lincoln," Taylor fumed. "Let the government send a man . . . with a few moral scruples—and I shall gladly give up all my pretensions and go home." Despite the teasing or the criticism of his associates, Clay was proud of his attractiveness to the fair ladies, and he preened himself for their approval.[16]

But while he was in the Russian capital, Clay was involved in one escapade which, socially, proved embarrassing—the Chautems affair. Eliza Leonard, a British subject living in St. Petersburg, had married Jean Chautems, a talented but impecunious Swiss chef. Somehow Clay met the family and admired their two daughters, the elder of whom, Leontine, was about fourteen years old. The parents had the misfortune to be imprisoned for debt, and Leontine went to Clay and asked his aid. The British secretary of legation, John S. Lumley, got Mme Chautems out of prison, but the family had nothing with which to face the rigors of the approaching winter. Clay and Lumley contributed enough to set Mme Chautems up in the boarding-house business, and Clay supervised the purchase of furniture and the renting of a house for her. She promised to repay the loan out of her income. Thus far, Clay's connection with the Chautems family was nothing more than generous charity. But he used his generosity as an excuse to continue visiting the family in its new home. Lonely and bored, he enjoyed the company of anyone who could speak English. He was particularly attracted to Leontine, and took her for drives in his carriage to the islands of the Neva. But he discovered that some of the furniture he had put in the house was missing and had been sold. Moreover, the woman never made any payments on her debt to him. His temper flared; he called the police to turn the woman out of the house, and he got a court decree to sell the furniture he had put into it. His action precipitated a blackmail scheme by Mme Chautems.[17]

To his embarrassment, Clay now perceived that his charity toward the Chautemses admitted of more than one interpretation. In his quest for informal companionship, he had provided the unscrupulous woman the opportunity to trap him. In a public statement, Mme Chautems declared that Clay had rented the house to serve as a trysting place, though she did not recognize it at the time. Since he seemed to be a benefactor, she went on, she had confidence in him until he took advantage of his position. He first attempted his immoral designs upon Leontine, she charged. When he took her in his carriage, he had a conversation with her, said her mother, which she, "in her innocence and simplicity," did not understand.

Failing in that "infamous" purpose with the daughter, Mme Chautems went on, Clay visited the mother when she was alone and sick in her bed, and attacked her "most brutally." Her virtue was saved, she said, only by the unexpected return of her daughter. Clay, "interrupted in his criminal intents," the woman concluded, then had them ejected from the house. She sued him in the Russian courts, but her case was thrown out because the defendant enjoyed ministerial immunity. Thereupon, Mme Chautems submitted her petition directly to the United States Congress.[18]

In the United States the Chautems petition did not arouse much attention. In April, 1867, Seward brought the matter before the cabinet, announcing that Clay was accused of licentiousness, seduction, refusal to pay his debts, and "pleading his representative character when sued." In Congress, the petition received the same treatment as did many others; it was referred to the House Committee on Foreign Affairs, and then to the State Department. Seward sent it to Clay with instructions to answer it fully.[19]

In the spring of 1867, Mme Chautems' allegations did not replace the Reconstruction Acts in the newspapers, but in Kentucky there were ribald reverberations to the petition. An anonymous pamphlet, purporting to be the work of "Timothy Bombshell and others," soon had readers chuckling over Clay's exploits. Entitled *A Synopsis of Forty Chapters Upon Clay, not to be found in any treatise on the Free Soils of the United-States of America heretofore published*, its perpetrators cleverly set themselves the task of solving the mystery of the "C. M." in Clay's name. In the beginning of his career, they said, the letters meant "Crying Mammiferous;" and at other stages in his life they stood for "Clamorous

Maniac;" "Callous Malignant;" "Cloddy Muddle;" and "Colossal Monstrosity." But after the Chautems affair, the pamphleteers joked, his name would only be "General Cohabit Misogamy Clay." There were many Kentuckians who joined in the raucous laugh.[20]

While Kentuckians ridiculed, Clay's friends in Russia hurried to his aid. The Russians, tolerant of his shortcomings, offered him the use of government agencies to contest the Chautems allegation. The secret police supplied him with affidavits to show that the woman was a recognized prostitute, and that her word was worthless. The police also affirmed that Clay's furniture had been located in the hands of a merchant who declared that the Chautemses had sold it to him. Prince Gortchacov told Clay to let his past life be his answer to the charges, and to pay no further attention to the scurrility. Sir Andrew Buchanan, the British ambassador who had handled the Chautems case, certified that Clay was blameless in the affair. Mme Chautems fled the empire to "escape her creditors," and the matter—in the Russian capital, at least—was soon forgotten.[21]

Clay's documents of the Chautems affair, which he spread before the world in his *Memoirs*, established his innocence, but they did not excuse his indiscretion in putting himself in such a questionable position. If, as his evidence suggested, the Chautems couple had a "very bad reputation with regard to their honor and morality," and were trying to "speculate upon the chastity" of their daughter, Clay should have had no dealings with them—at least, not for the public to see. The Chautems affair was the most lurid episode in Clay's colorful career as a diplomat.[22]

But in 1866 Clay did more than involve himself in scandalous escapades. That year, an unsuccessful attempt had been made upon the life of the Czar Liberator. The United States Congress, mourning over the recent assassination of President

Lincoln, passed resolutions congratulating the Czar upon his escape, and appointed Captain Gustavus V. Fox, former Assistant Secretary of the Navy, to deliver them in person. With an ironclad monitor, the first such vessel to appear in North European waters, and two wooden vessels, Fox made his pilgrimage to the Russian capital. With all the pomp he could muster, Alexander II received the special American envoys, and Clay busied himself arranging the details. Secretary Curtin proved invaluable to the Minister, serving as interpreter for the group and also as its escort on a tour of inland Russian cities.[23]

Clay was still well pleased with his secretary, but soon his uncontrollable jealousy appeared. In 1866, Curtin visited Moscow, and the merchants there planned a banquet in his honor. Just before the affair was to take place, Clay, worried about his ministerial popularity, suddenly appeared and had to be included on the program. Years later, when both Clay and Curtin wrote their memoirs, each of them recalled that he had been the chief attraction at the Moscow banquet: Clay boasted that his anti-English, home-industry speech, in which he compared Russia to the American South, won the acclaim of his mercantile hosts; Curtin claimed that his stirring toast to Moscow delivered in the Russian language aroused more sentiment. But as yet there was no evidence of hostility between them. In reporting the affair to Washington, Clay highly commended his secretary. "He is a great acquisition to this legation," he said, "and deserves well of the country." Clay even suggested that the United States consulate in Moscow be closed, and the salary given to Curtin, who was "*starving.*" [24]

But a few months later, at another Moscow banquet, Clay angrily accused his secretary of trying to steal the show. After that, Clay's bitterness toward the young man mounted. He forgot the compliments he had earlier paid Curtin; he put out of his mind even the manner in which the young man had entered the legation staff. Clay accused the secretary of turning the Russians against him in order to seize his position. He charged that Curtin had assisted the Chautems woman to spread "calumny" about him, and he peremptorily ordered the secretary out of the legation.

Taken aback by Clay's sudden venom, Curtin never again entered the building. Since the Secretary of State kept the young man on the salary list until 1869, Clay complained that Curtin represented but another of Seward's efforts to pillory him. "He smuggled in upon me the Jesuit and drunkard Jeremiah Curtin," Clay declared to Senator Henry Wilson of Massachusetts, "whom he has used as a spy and calumniator against me ever since he has been here." As Seward's tool, Clay went on, Curtin was the cause of all his woes. "He was the aider and abetter of the bawdy-house keepers, J. and Eliza Chautems, in sending on the libelous petition to Congress against me. . . ." To Seward, Clay fumed that "no honorable man would ask me to associate with an ingrate, a calumniator, and a proven swindler."

Clay had practiced his invective for many years, and with Curtin his scurrility was at its height. He loaded his diplomatic reports with harsh words about his secretary of legation, until one of the State Department clerks complained. "Has a man simply because he happens to be a minister," he asked the Secretary of State, "a right to put a number at the top of a sheet, and then fill it with scurrility and obscenity and require you to preserve it in the archives?" Clay's jealous hatred for a possible competitor had overcome his self-control, and he hurled imprecations at Curtin.[25]

Clay wrote also to many politicians in the United States, reiterating his charges against his secretary. "Curtin is the most abandoned scoundrel I ever met of his age," he declared to Schuyler Colfax. "It would take a volume to tell you all his scoundrelism . . . the whole course of the poor devil seemed to have been inspired by his hope of being *chargé* here." Clay denounced Curtin as an "habitual drunkard and also an acknowledged hypocrite." But to Seward, Clay's story was different. Curtin was the "*protégé* of Mr. Sumner, and shows his hatred of me," he declared. The cause of the dispute, he told Seward, was a toast at the Moscow banquet which omitted Seward's name but included Curtin's. After the toast, Curtin "had his claquers ready who . . . tossed him up—to the disgust of all the gentlemen present." This was, Clay continued, "but one of a thousand of his pieces of impudence." Obviously the secretary was too popular for the Minister.[26]

When Seward called upon Clay to give specific charges against the secretary, other than jealous hatred, the Minister presented evidence that Curtin had failed to pay his debts. The young man responded, as Clay himself had reported, that living in St. Petersburg was too expensive for his salary, but that as soon as possible he had repaid his creditors. But he overlooked one bill for £2 4*s*., which the obsessed Clay heralded as proof of Curtin's perfidy. In view of his own extended indebtedness, it revealed a lack of understanding that Clay pounced upon a small obligation of short duration (and apparently an honest oversight at that) to besmirch Curtin's character. For Clay to magnify so petty a debt into a continuous tirade indicated the extent of his jealousy. For him to hold Curtin guilty of crimes because of his religion suggested that Clay was hard-pressed for a cause to attack the young man. But Clay's impassioned tirade suggested also that he had other reasons for denouncing his secretary.[27]

There was an important business deal which led Clay to seek the removal of Curtin. He wanted a certain M. D. Landon as his secretary, because Landon was related to Robert Williams, the "great American railroad man of Moscow." Having profited from the Russian-American Telegraph Company, Clay now saw an opportunity to make money on Russian railroad construction. Already he was casting about for an American accomplice to raise capital for the scheme. "The Russian Government are very anxious to complete their railroad system and will grant very liberal terms," he confided to John A. Andrew, Civil War governor of Massachusetts. Clay invited Andrew to join him in forming an investment company to lend capital for the railway expansion. "I have the confidence of operators here as well as of the government," he said, "and if you think proper to try your hand in the matter, we can make a large fortune—say a few millions of dollars. . . ." With such a vision before him, Clay desired to rid himself of Curtin, who had no influential connections, in favor of the railroader's relative. Even as minister, Clay was primarily concerned with capital investment and industrial development.[28]

The railroad scheme fell through, but a business matter more important to the public did not. In the spring of 1867, a treaty was ratified which provided for the purchase of Russian America by the United States. Clay claimed credit for the acquisition. When he was attacking the Secretary of State, he made the accusation that "Seward has also attempted to appropriate all the honor . . . for the Alaska and other concessions of the Ra. [Russian] government—whereas he is known here as the enemy of the Foreign Department, and has no influence whatever—and all the favors of Russia are due only to me." Years after that immodest avowal, Clay was still trying to prove that his influence had been responsible for the Alaska purchase. When all his other accomplishments were forgotten, he told the students at Berea College in 1895, posterity would remember him as the author of Alaska annexation. All he wanted upon his monument, he said, was his name, and the single word, "ALASKA."[29]

Despite his claims, Clay had played only a small part in the purchase of Alaska. The discussion of it took place in Washington between Seward and the Russian Minister,

"Baron" de Stoeckl, and when the treaty was worked out, Seward obligingly sent Clay a copy. The Minister in St. Petersburg had had some dealings with the business interests which desired the territory, and through that connection he claimed credit for the purchase.

During the 1860's it became evident that fur-trading rights to Russian America, or Alaska, would soon be open. Because of declining profits, the Russian-American Fur Company, which had had a monopoly of the trade, did not desire an extension of its charter when it expired in 1862. Part of its rights, however, had been granted to the Hudson's Bay Company (unpopular in Russia), in a sublease which ran until 1867. In California, a group of Americans formed a company and laid plans to receive those rights. Because of his success in urging the claims of the telegraph company, Clay worked with the American company, and he reported optimistic hopes for getting the charter. On his way back to Russia in 1863, Clay met Robert J. Walker, Polk's Secretary of the Treasury, territorial governor of Kansas, and confirmed expansionist, who was on a special mission in Europe. "I impressed upon Walker the importance of the ownership of the western coast of the Pacific," Clay reported to Seward, "in connection with the vast trade which was springing up with China, Japan, and the western islands." [30] Actually, Clay's arguments were not necessary to convince Walker, already in favor of acquiring as much West Coast territory as possible.

The Russians were contemplating selling Alaska outright, rather than merely granting a charter for hunting rights. The Czar's representative in Washington, de Stoeckl, considered the territory a potential breeder of trouble between the two countries, and recommended the sale. In 1866 he received permission to offer it to the United States. He bargained with Seward and succeeded in getting the price raised from five to seven million dollars, and the deal was made. So anxious was Seward to close the matter that he kept weary State Department clerks at work all night ironing out the details. Getting the treaty ratified by the Senate, and even more important, getting the House of Representatives to appropriate the necessary funds, were problems which would cause the Secretary much worry and much tactful wire-pulling. While the House wrangled over the treaty, Clay wrote from St. Petersburg that the Russians were expressing the hope that the cession would lead ultimately to the expulsion of England from the Pacific. But despite his attempts to influence the lawmakers, Clay had little part in the Alaska deal.[31]

After the annexation, the remainder of Clay's term in St. Petersburg was uneventful, and he prudently managed to stay out of trouble. His career as a diplomat was over, and he referred his ministry to the mercies of posterity. "As to my diplomacy, I leave that to history," he said, "What reason was there why Russia should stand by us, when other monarchies desired to destroy us? . . . Who shall say then how much all this is owing to myself?"

In keeping Russia sympathetic to the United States, Clay's ministry was a success, in spite of his diplomatic blundering. He was a popular representative, and well liked by many Russians, but his hot temper and his eccentric social manner marred his ministry. After Clay returned home, the Russian Minister, Catacazy, got into difficulties with the State Department. Someone told the Secretary of State, Hamilton Fish, that "after all, I have a notion you ought to be pretty patient with Catacazy when we reflect how long St. Petersburg bore with Cassius Clay." Clay was able to remain in Russia for more than six years only because the Russians were tolerant of his manifold aberrations.[32]

CHAPTER XV

INDIAN SUMMER

IN 1863, when Cassius M. Clay left the United States with his high hopes for prominent office unfulfilled he had returned to Russia and become a Lincoln supporter. "If 'Uncle Abe' desires it," he confided to Salmon P. Chase in 1863, "I rather think he deserves another term." Cassius was far from the center of American political activity, and until his home-coming in 1869 he played only a commentator's role. In Russia he had acted as though the career for which he had been so ambitious had come to an end.[1]

Back in America he soon discovered that the party was divided by factional disputes, and that there was no politician of Lincoln's calibre to control its divergent elements. There was strong opposition within the party to the Radical program of reconstruction. Behind the dissatisfaction was sufficient force to create a revolt, and in it he might yet attain the position which the regular party would not provide. Clay therefore became a critic of the party leadership. "New men have come into the control of the Republican Party, who never had any sympathy with the original Republicans," he protested, "who, assuming our watchwords, seizing our ships, overthrowing our commanders, and sailing under false colors, have since waged a more destructive war against the glorious principles of republicanism than the rebels themselves."

The Radicals, Clay charged, had declared war upon the states: "The rebels attempted the overthrow of states by dissolution; the radicals have accomplished the same by *centralization*." Making that charge the basis of his campaign, Clay re-entered American politics and sought to become the spokesman for the anti-administration malcontents. For Cassius Clay, who had spent the summertime of his life in the Baltic cold, and for whom an extended winter of retirement lay ahead, it was Indian summer. He prepared to make the most of his new opportunity.[2]

Once more, Clay had seized upon a potentially popular issue. He was not alone in criticizing the administration of the defeated South. Though he was among the first to declare himself a "Liberal Republican," there were others ready to take such a step. Underlying the military action against the Confederate States of America was the economic conflict between the interests of the industrial, mill-owning North and those of the agrarian, plantation-owning South. After the enemy had been defeated, and after a system conducive to business prosperity had been established in the North, the next phase was the control of southern economy. To accomplish that purpose, the Radicals first employed coercion by military occupation, which carried with it the establishment by force of Negro political equality. That course so aroused stubborn southerners that a state of near-anarchy resulted, and in it, investments were endangered. Many moderate northerners, unconcerned about the Negro, deplored the Radical policies which engendered strife rather than establishing peace and order. In 1869, there were many influential men who grumbled at the administration's policy. Cassius Clay moved to capitalize upon that discontent.[3]

Immediately upon his return from Russia, Clay joined the rebellious Republicans in denouncing the Radical policies. For years he had been registering his disapproval of a vindictive, coercive treatment of defeated southerners. In 1863, he had urged the acceptance of a more liberal attitude. "Here is a great rebellion. We must either destroy all engaged—or forgive them," he told War Secretary Edwin M. Stanton. "I say our policy is to execute the unrepentant rebels: and forgive the repentant. It is just—it is good policy." And to his friend Chase he repudiated Senator Charles Sumner's thesis that the states of the South had committed suicide and had reverted to territorial status by seceding. "After the rebels are disarmed, the states and their rights *revive*, except so far as individuals may be affected by legal procedure." Holding such views, it was easy for Clay to find fault with postwar Radicalism.[4]

Clay had also long been a critic of the administration. He had believed himself persecuted by the Secretary of State, William H. Seward, and when Hamilton Fish succeeded to that office, Clay charged that Fish was a dependent tool of Seward, and that he continued Seward's unprovoked insults. Clay was therefore arrayed against both the philosophy and the personnel of the Grant administration. To dramatize his allegation that party leaders were apostates to "original republicanism," he made use of a cause lying ready to hand. That issue was the Cuban Revolt.

In 1868, patriots on the Spanish-held island had launched a rebellion for independence in the hope of obtaining assistance from the American mainland. With the strain of Reconstruction and the prevalent war-weariness following the four-year struggle at home, the administration, and in particular Secretary of State Hamilton Fish, hesitated to become involved in a struggle with Spain. But Cassius Clay had no such restraints, and, to embarrass the administration, he promptly plunged into the troubled waters of Cuban affairs. It was fundamentally the same issue he had always preached—emancipation—and for the same purpose—political office.

Without even taking time to visit his family, whom he had not seen for over six years, Clay planned an active campaign for Cuban relief. He expressed horror at the official apathy which met the gory atrocity stories emanating from the tortured island. He even charged that the administration favored the Spanish over the oppressed natives. "When I arrived in New York, in 1869, Spanish gunboats were fitted out in New York harbor," he recalled later, "whilst the Cuban masters and their liberated slaves were spied out, and all their ships and material confiscated." Decrying such favoritism, Clay called upon his veteran abolition colleagues to join him in a protest movement. When he tried to speak in Cooper Institute on behalf of the rebels, he was rudely received by the New York "Custom-house dependents," the Grant henchmen, and could not make himself heard. The behavior of these few hotheaded urban ruffians played into Clay's hands; he used that incident as ammunition to fire at the Republican leadership. "Thus I, who had never failed to secure, in the slave States, a hearing was, in the free city of New York, silenced!" he complained, insinuating that the Radicals were more inimical to liberty than were the southern aristocrats.[5]

Clay now considered himself at war with the administration. The city roughs had helped him to draw the line between his position and that of the Radicals. Despite the noisy opposition, the Cooper Institute gathering accepted Clay's resolutions, and he was elected president of the newly organized Cuban Charitable Aid Society. It was the first step toward the creation of an independent political party to oppose Radical Reconstruction. Under the guise of organizing philanthropic aid for an oppressed people, Clay attempted to form a standard around which moderates could rally. In the statement of purpose which he published in the name of the society, he declared that he wanted to force the administration to change its policy toward the Cubans. "Our purpose," he announced, " is to arouse and concentrate the moral support of the nation in behalf of the recognition—by the general government—of the Belligerency and Independence of Cuba." [6]

In the Cuban Aid Society, Clay had distinguished company. Horace Greeley, another disaffected, crusading Republican, was a vice-president of the society, and Charles A. Dana, erudite editor of the New York *Sun*, was its treasurer. The movement now had an organization and a press, and Clay used it to determine the extent of anti-Radical opinion within the Republican Party. In expressing his disapproval of administration neutrality toward the Cubans, Clay made the issue a test of allegiance. Any politician who would join in castigating the administration on that question might also serve in the ranks of a protest party.

Some leading Republicans, like Senator Benjamin F. Wade of Ohio, concurred in Clay's criticism of the Grant-Fish foreign policy. "I am astonished at the apparent indifference of the great Republican Party to the fate of the people of Cuba," Wade told Clay. "Are they indeed weary in well-doing, or do they still favor that timorous, halting, hesitating policy, which added more than half to the blood and treasure in conquering our own rebellion . . . ?" While Wade sympathized with Clay's Cuban Society and agreed to serve as vice-president for Ohio, other prominent party leaders rejected the manifesto. Governor John M. Palmer of Illinois, although he later became an outspoken critic of the Radicals, was at first cautious and refused to participate in the Cuban affair.[7]

Clay's new emancipation program attracted little attention. The society collected less than two thousand dollars, of which about half went for relief to the Cuban *junta*, and the rest to organizational overhead. Despite its poor showing, it served its purpose, and Clay was satisfied with its effect. He protested loudly that the opposition of President Grant and Secretary Fish had effectively curbed popular demand for Cuban recognition. The Republican Party, which had come into being to defeat American slavery, Clay said, had deviated from its noble purpose of liberation. The Cuban Aid Society had proven that criticism of the Grant administration and its policies was popular. For Clay's political ambitions, the society was a success.

But Clay could not immediately follow up his initial victory. He had personal problems which demanded his attention. His marriage, which had been strained for many years, was on the verge of dissolution. Mary Jane Clay was among those Kentuckians who believed the Chautems blackmail story, and she no longer considered Cassius her husband. "When Seward calumniated me, she wrote a letter, not trusting me, but believing the Chautems' scandal," he said, "at which I was indignant and thus closed our correspondence." When he returned to Kentucky, his wife treated him as a stranger and moved him into a separate room. The weather turned cold, and there was no fireplace in his room. He told of enduring cold "so intense that icicles froze on my beard." Soon after that indignity, Mary Jane called him in for a discussion of his financial affairs. In the course of the discussion, he alleged, she lost her temper and poured out angry words at him. Infuriated, she delivered imprecations, as Cassius reported it, "upon my devoted head like a deluge." At first, Cassius recalled, he was angry. "After I had married her," he declared, "my love for her was pure and devoted, *and it was she who made the first breach upon the marriage duties*." But when he saw her fury, he made no response. "The last touch of love had vanished," he said, and he let her finish the tirade. Then he politely bade her good-night, and left her. His marriage was at an end.

On two occasions after the separation Mrs. Clay humbled herself in an effort to bring about a reconciliation. She sent him a message that she loved him as much as ever, but Cassius met it with a tight-lipped silence. When he heard that she intended to move back into White Hall he abandoned it to her, leaving it vacant. With her offers rejected, Mrs. Clay withdrew to Lexington and established a separate residence. In 1878, after five years of separation, Clay sued his wife for divorce, and she did not contest the suit. On February 7, 1878, he was legally "restored to all the rights and privileges of an unmarried man." After the judicial settlement, Mrs. Clay lived in Lexington until her death, at the age of 86, on April 29, 1900. Because of Cassius' stubborn unforgiving spirit, both he and Mary Jane were lonely for the last twenty-five years of their lives.[8]

But Clay lost little time bemoaning his broken marriage. He considered that he had a new chance for high office, and he enthusiastically entered national politics. To separate himself from the Radical-dominated Republican Party, he joined the Democrats; but he did not make an overt change for several years. In 1870 the Democratic Party was still tainted with secession and was therefore unpopular. Until it became acceptable to avow himself a Democrat, Clay helped to organize the amorphous protest-movement by working with independents. "The Democrats were beaten in the war, and were powerless in politics," he explained later. "I thought that the way to help them was to start an independent candidate, which, if they were wise, they would support with or without a convention nomination." Clay accordingly became a pioneer among the Liberal Republicans.[9]

With that as his purpose, and with the ever-present hope of reward, Clay continued his attacks upon the Grant administration. "I think you will . . . find out that it is utterly impossible to keep the Republican Party in power with Grant and the imbecile sycophant Fish as *regent*," he told Massachusetts Senator Henry Wilson. "I wish you would join yourself with those who desire a liberal course towards the South," he urged. "I feel as sure as I live that the policy of *repression* will fail. . . ." Clay claimed that his entreaties convinced many of the old-line Republicans that a more moderate policy would bring the desired results. Such men as Sumner, Greeley, Chase, Julian, and Lyman Trumbull, he said, agreed with him that the Radical program was a mistake, and hindered the effort to reconcile the South with northern economy. Clay also asserted that all "careful thinkers" perceived that the policy of putting the Saxon race under Negro rule would fail.[10]

After private correspondence with old-school Republicans had convinced him that he would receive their support, Clay began denouncing the Grant leadership openly. In the summer of 1871 he advocated Horace Greeley for the presidential nomination. On July 4, at Lexington, he was back on the stump in his home town with a bitter attack upon the military rule of the South and the coercion policy. Clay wanted southerners to plan their own economic development, and he said that the Radical scheme of inciting the blacks against the whites prevented it. "The interests of the blacks and the old masters are the same," he said, "and if they are just they would regain their former power not by revolution, but by education and the development of the unequalled resources of the South."

Clay appealed for southern support of the Liberal movement and held out the promise of restored local control. Repressive rule from Washington, illustrated by the Force Bills of 1871, had taken the place of the aristocratic agricultural party in hindering the industrial development of the South. Clay urged a change in the national government. He favored a candidate who came with the olive branch, rather than the sword—Horace Greeley, rather than Ulysses S. Grant. The first step in Clay's final drive for power was to break the Radicals by restoring the southern states to local control.[11]

In pursuing that course, Clay became an ardent champion of states' rights. Early in 1872 he published a proclamation justifying the Liberal Republican Party and his own aspirations for office. "The Constitution was based upon the vital integrity of the states," he began, "and their unhappy overthrow was not necessary to the suppression of the rebellion or to the liberation of the slaves, or to any legitimate purpose for which the war was waged." But in addition to the destruction of the states, the "proscriptive rule of the Grant party," Clay charged, had an even more serious consequence: it left the North and the South still hostile and the subjected states in anarchy. "It was the disfranchisement of the leading minds of the South, and the fatal attempt to subject the Saxon race . . . to the minority of the African freedmen, which bred the foulest excrescence of slavery—the Ku-Klux clans."

Clay condemned the Radicals for disrupting order in the South, and he appealed for the support of southern Democrats to the Liberal movement, still promising local control. "If the Democrats are wise and just," he admonished them, "the blacks, five millions of the South, will vote with the South; not because they are Democrats or Republicans, but because they are Southerners."

In the interests of his own aspirations, Clay posed as a southern patriot. "As a southern man . . . ," he explained, "I have devoted my life to the overthrow of slavery—because it was unjust to the black, and a cause of weakness to the white." Now that slavery was dead, he was still sympathetic to the South. "I resist with the same earnestness . . . the attempt of the Grant conspirators to subjugate the South," he declared. "I denounce the attempt to weaken us by a studied policy of barbarizing us, by the corrupt and irresponsible rule of men from the North. . . ." Once more Clay spoke for southern sentiment, and his motive was the same as before. "We want new men, with new sentiments of good will," he told the southerners, speaking of Horace Greeley. "We want men who . . . turn their backs upon rebellion and secession, and look to the Union only for safety and happiness. . . ." [12]

Clay carried his support of Greeley into the National Liberal Republican Convention. In May, 1872, at Cincinnati, a confused assemblage of disgruntled politicians, old abolitionists, and hard-money enthusiasts gathered to name candidates to oppose Grant and the Radicals. As leader of the Kentucky delegation, Clay supported Greeley, who received the presidential nomination. Clay claimed that he was responsible for securing Greeley's nomination over Massachusetts' scholar and diplomat, Charles Francis Adams, and in the campaign which followed, Clay vigorously campaigned against the Republican ticket.[13]

In his campaign speeches he continued to condemn Radical rule in the South. He was shocked upon his return to Kentucky, he said, to see state sovereignty overruled, and troops patrolling the state which had made more sacrifices for the Union than any other state. "Why? Because other men had interests in maintaining the Union. They had pecuniary interests as well as patriotic motives. . . ." Clay denounced the carpetbaggers, and in effect he abandoned the Republican Party. Its mission had been accomplished, he insisted, and it no longer served a useful purpose. "It was no part of the Republican program . . . that the blacks should be placed above the whites."

Clay also grieved over the state-destroying policies of the Radicals. "Let us save the States; let us save the South;" he said, "and by saving the States and saving the South, we save the Union and the liberties of the . . . Constitution." Southern rebels had gone too far in seceding from the Union, Clay acknowledged; but the Radicals, "in ignoring all the rights of the States, and engrossing the whole power of the Government in the National Administration," had gone to the opposite extreme. As he had opposed the one mistake, so he also fought the other. To Negroes Clay gave the advice that they cease their enmity toward their former masters, and assured them that they would not be returned to slavery. "Do you owe an allegiance to the present administration?"

he would ask them. "No; you owe it to those who fought your battles. You owe no gratitude to Useless S. Grant; he never voted for you in his life." [14]

Clay also called for peace between the sections, which the Radical program did not provide. Though it was seven years after the end of the war, he said, "we hear to-day, the same battle-cry that was begun in 1860." He told an Ohio audience that it was impossible to maintain the republic with "fifteen, sixteen, or seventeen States of the South in a chronic state of enmity to the United States." That was the reason, he explained, that he campaigned for the Liberals. A more moderate policy might ease southern qualms and allow peaceable northern investments.[15]

Despite Clay's efforts, Grant was re-elected. But the Kentuckian did not regret his work with the Liberals. The party had served its purpose: it had united resistance to Radical Reconstruction, and it had prepared the way for a Democratic opposition which would eventually end military occupation. "Greeley was beaten," Clay said later, "but the Democratic or opposition party was placed upon the road to victory." In the Indian summer of his political career, Clay was carefully cultivating the crop he hoped to reap: nomination in 1876.[16]

Until then, Clay went into semiretirement at White Hall. "I'm leading a rather solitary life," he reported in 1874, "and am much at home, and devoted all the more on that account to flowers and fruits." His farm and his garden occupied much of his time, and he read for many hours every day—Shakespeare and Milton, as well as history and biography. Visitors to White Hall were impressed by Clay's dynamic personality and his urbane, enlightened conversation. A little above medium height, with stocky, powerful limbs, his face covered with a thick, iron-gray beard and his hair silvery gray, Clay had lost none of his charm in his retirement. He was lonely in the big house but he lived in simple luxury among his flowers, his orchards, and his books. He kept a close watch upon public affairs and maintained a large correspondence.[17]

In May, 1875, Clay brought his retirement to an end, and began the final step in his campaign for high office. He openly joined the Democratic Party at its state convention in Frankfort, thus completing his break with the Republicans. An important ex-Republican, he attracted much attention by his shift, and his reasons were widely circulated as campaign material. "There is but one great issue between the Republicans and the Democrats," he explained, "but that issue is the most important that ever interested the human race . . . whether man is capable of self-government." The people of intelligence and moral worth—the propertied classes—ought to rule a country, he said; and in the southern states that was no longer the case. "When the South . . . laid down her arms, she should have been restored at once to self-government," he proclaimed. "The attempt to rule the eleven States . . . from Washington, was only possible by overthrowing constitutional government and becoming a central despotism." Grant's tyranny, "directed towards the Southern people, with whom I am identified," Clay explained, "drove me into the Democratic Party. . . ."

As a recognized Democrat and an advisor to the party, he urged the nomination in 1876 of a "straight-out" Democrat for the presidential candidate, with the second office going to "a liberal in the South." Clay's ambition to be vice-president would not let him rest, and some of his acquaintances agreed that he now had a chance. "If an Eastern man is selected for the first place," the Kentucky secretary of state told Clay's nephew, "your Uncle might stand a fair chance for the second, unless the point would turn on a financial compromise and the ticket be Tilden and Pendleton." [18]

Having declared war upon the Radical Republicans and

disclosed his own aspirations, Clay set out to garner publicity for himself by making speeches. He reiterated his interest in southern problems, expostulating against Radical despotism and the anarchy which it produced. He simplified the Democratic platform to the one issue of state control in the South. To head off the Pendleton movement, he tried to minimize the financial issue which faced the country. "What we need is not more currency, but more confidence in State values and good government," he told his fellow Madison Countians. "I want to go into the canvass of 1876 upon the true issue—the self-government of the people of these States. . . ." [19]

Clay delivered numerous speeches for the Democrats, and he participated in state elections in such widely diverse areas as Ohio and Mississippi, to demonstrate his vote-getting appeal. In Mississippi, where Clay made several appearances, the voters overthrew Adelbert Ames, of Maine, and the carpetbag regime, and the Kentuckian claimed credit for the "oasis in the gloomy desert of Democratic defeats." His success in the deep South inspired him to proclaim his candidacy. He wrote bundles of letters to prominent anti-Grant politicians, begging their support, and pointing to his ability to win votes. "My name will be presented to the Demo. Nat'l Convention next year for vice president," he would say. "North will get first place, and second naturally goes to South. . . . I ask your favorable consideration." The time had come when Clay hoped to gather in the harvest for which he had worked.[20]

But his elaborate preparations bore bitter fruit. When the Democratic convention met, his candidacy was not strong enough to sway the party. Samuel J. Tilden, a "straight-out" Democrat of New York, won the presidential nomination, as Clay had anticipated, but for the second place the delegates chose Thomas A. Hendricks of Indiana. As he had done so many times before, Clay had to swallow his disappointment and campaign for another candidate. Once more he expected a cabinet appointment as his reward. "I hasten to assure you," one of his Kentucky friends told him, "that nothing could afford me more satisfaction than the opportunity I now have of recommending your appointment to a position in Mr. Tilden's cabinet."

But though Clay received support for the coveted position of Secretary of War, Tilden did not become President. The Democratic candidate received a popular majority; he did not, however, get enough uncontested electoral votes to secure election. In South Carolina, Florida, and Louisiana, each party claimed to have carried the states, and each claimed the nineteen electoral votes which those states cast. In the dispute, a special electoral commission appointed to determine the validity of the votes in those three states gave them all to the Republican candidate, Rutherford B. Hayes, making him President. But an elaborate compromise behind the scenes prevented another civil war and guaranteed the victory to the Republicans.[21]

Despite the peaceful settlement of the dispute, many observers expected hostilities over the election. "I am not at all sure that the present complications will end without a fight," one of Clay's acquaintances told him. "If the government can be seized and held by a gang of unprincipled political buccaneers against a popular majority of over 300,000 . . . there will be some new and startling questions in American politics before the four years shall elapse. . . ." There was no conflict, but Clay was disgusted with the manipulations which prevented his reward. Tilden was elected in 1876, he categorically reported later. "It was no fault of mine that he was by Democratic treachery and cowardice, and Republican fraud and bulldozing, not allowed to take the place to which the people . . . had assigned him." But regardless of the arrangements which made it possible, Hayes's election brought Clay's campaign to an end.[22]

Once more Clay's aspirations to high office had come to

failure. Because of his shift in political allegiance, he held the unique distinction of having made a serious effort for nomination in both major parties, and of having failed in both. In 1860 his southern background and his extremism deterred his campaign, and in 1876 his long career as a Republican detracted from his availability.

Although he did not receive the reward he craved, his effort nevertheless revealed his political aptitude. He had guessed right in 1876, just as he had guessed right in the 1850's, but circumstances beyond his control worked against him in both cases. Of all the hard-luck candidates in American politics, Cassius Clay could take his place alongside his kinsman Henry Clay, at the forefront. And with the failure of his campaign among the Democrats, Clay's Indian summer was over, and winter was at hand.

CHAPTER XVI

THE LION OF WHITE HALL

WITH the defeat of the Democratic Party in 1876, Cassius M. Clay declared his career at an end. The first act of the new President, Rutherford B. Hayes, was to terminate military occupation of southern states, and Clay claimed that his part in the campaign had been a significant factor in that action. "Here the two great acts in the Political Drama, in which I have borne a soldier's part, cease," he said a few years later. "The first was the freedom of the blacks . . . ; the second was the restoration of the States to their original sovereignty." Declaring his efforts successful, Clay retired into lonely leisure at his Madison County estate. But though he withdrew, he did not escape public notice. He continued to play the exciting game of politics, in the capacity of elder statesman, and he indulged in characteristic antics which aroused attention. It was during his retirement that Cassius Clay won the sobriquet, "Lion of White Hall."[1]

In the years after 1876, Clay, whose entire career had been colorful, became the subject of much folklore. As he grew older, a kind of mental illness appeared, which drove him to wild fears and dark forebodings. His tortured mind finally broke under the strain and the loneliness, and he mistrusted even his own family. To forestall his imagined enemies he turned his mansion into a fortress and diligently guarded it.

The final years of Cassius Clay's life, like the stormy career which had preceded them, brought him no peace. Haunted by insane fears, lonely to the point of desperation, and childishly senile, he was a pathetic caricature of his younger self. Despite all his handicaps, however, Clay still possessed enough of his earlier spirit to earn the title of "Lion."

White Hall was a fitting lair for such a personality. The house, remodelled while Clay was in Russia, was now a luxurious mansion capable of serving as a fort. Clay had had the house constructed of specially prepared brick knit together with mortar of a high lime content, and with no wood exposed except in the window frames. Inside the defensive shell, however, the house lost its forbidding atmosphere. In the entry hall, so large that "you could turn a wagon load of hay about in it without touching the walls," (a visitor politely exaggerated), were niches for statuary. Clay possessed busts of Henry Clay and Horace Greeley by the Kentucky sculptor Joel T. Hart, and he also displayed portrait paintings of the Russian Czar and Czarina and the Prime Minister, Prince Gortchakov. To the right of the hall were parlors with walls twenty feet high and marble pillars supporting the ceiling. On the left of the hall was an oval library, and back of that was the dining room. From the entry a graceful staircase curved upward to the second floor, where there was another enormous hall, and along it spacious bedrooms, with walls fifteen feet high, and for each bedroom a large private dressing room.[2]

In addition to the ordinary appointments of the house, Clay had made some unusual installations. On the roof of one of the porticoes was a fish pond visible only from the upper stories. An inside bathtub, for which the servants took water upstairs, was built into a small room; and in closets on the second floor were two commodes. Clay furnished the house richly, with carpets, chandeliers, mirrors, vases, and antique furniture more than a century old. His library was extensive.

He boasted "most of the first works of all time." With sturdy walls, spacious living quarters, and luxurious fittings, White Hall was a magnificent mansion.[3]

Among such comforts Cassius Clay lived out the lonely years of his retirement. Although he had never been an enthusiastic farmer, he now gave close attention to his garden. Fruits, vegetables, and grains grew profusely, and Clay loved to point out to visitors that everything upon his table, except the pepper, salt, and coffee, was produced on the farm. He was fond of mutton and was proud of his herd of purebred Southdown sheep. But though he supplied his table well, he was not a heavy eater. He attributed his longevity to his temperance in eating, and he loved to relate that he always concluded his meals while still hungry. As a result he remained lithe and stocky, and his eyes retained their brightness. Only his gray hair and his beard revealed his age: as he poetically described it, "The frost which never melts settles upon my locks!" But the years did not at first detract from Clay's personal appeal. His conversation was brilliant and compelling, and those who met him never forgot his fascinating personality.[4]

In his retirement the old man maintained an active schedule. His day began at 8 A.M. with a "morning nip of tansy bitters . . . as a simple tonic," followed by breakfast. Then came the bath. "I bathe more or less every morning in cold water," he explained, and he also boasted that he "never used tobacco in any shape, and spirits only very sparingly." His mornings Clay devoted to reading under the trees on the lawn if the weather permitted; or if not, he would move into a glassed porch. After a light lunch he would nap in the open air upon an iron lounge without pillows or cushions. "I want the air to circulate all around me," he said. "I am fleshy enough to dispense with cushions, and they attract flies. You see I have no flies around here . . . they are my abomination."[5]

But it was loneliness which bothered Clay. "He lives almost a hermit in this big house . . . ," a visitor reported. Cassius tried to convince himself that he preferred to live alone. "No baby to break things," he jocularly pointed out; "I will soon have a sign painted . . . 'Let no woman ever enter here!' " Brave attempts at humor, however, did not conceal his desire for companionship. In the absence of human company, Cassius surrounded himself with friendly animals. When he read under a tree, he would hang a bird cage from a branch. "I sought companionship with the flowers and trees and shrubs. . . . . I gathered about me dogs . . . and pigeons and barn-fowls, and the mute fishes. . . ." He erected a crumb-box near his window to attract wild birds. "But at night I was left all the more alone," he sadly mused, "till I often opened the shutters that the bats should enter . . . and their fluttering—life—life—was a pleasure to me." [6]

Clay did not submit tamely to the fate which punished him by abandonment and loneliness. Characteristically, he sought something to fight. "Was it of God, or of man?" he would angrily demand of the stars. "If of man, then will I contend with man—I will assert my eternal defense! And, if from Fate, then will I 'wage with fortune an eternal war!' " But he had met an enemy which he could not master by brute force, and he found no solace in his retirement. "I, who had sacrificed all to men, was by men left to myself alone," he bitterly complained.[7]

In desperation Clay made plans to secure companionship. He remembered the sympathy and love he had received from an unnamed Russian woman. "In a distant land . . . was one spark of eternal life, which . . . spoke to me in words which I could well understand," he said. He began to dream about his St. Petersburg mistress. "Day by day that one image—that one voice which for so many years in a strange land I had listened to as the sweetest music—gathered into vividness." To Clay, determined to endure loneliness no longer, the memory brought hope. In 1866 a son had been born to the woman, and Clay made no effort to deny his paternity. The lonely man sent for his child. Brought to White Hall, the four-year-old boy was legally adopted by Clay, and his name changed to Launey Clay, "by permission of his nominal parents." Clay knew there would be criticism and gossip, but for the sake of companionship he prepared to face it. "Having made up my mind as to my highest duty," he said, "I calmly shouldered all the responsibility for my action. . . ." As he expected, there was much tongue-wagging over the Russian boy. "You see an early champion of freedom walking about boastfully with a bastard son, imported like an Arabian cross-horse, and swearing at his family," one editor boldly remarked.[8]

But it was not only satirical journalists who attacked the young Launey. Clay, busy with his own affairs, paid little attention to the housekeeping chores, which he detailed to the servants, a Negro couple named White. Having assigned Launey to the care of a nurse, he considered his responsibility fulfilled. But he was alarmed when the boy became pale and listless and would drop things from his half-paralyzed hands. When Clay questioned the nurse, she explained that Launey ate dirt and was pining away from homesickness. Cassius began to watch the boy more closely but suspected nothing unusual until he intercepted a letter from Perry White, a son of the servant couple, to his mother. Then Clay learned that the servants had been systematically poisoning Launey with arsenic. Promptly Clay's face clouded, and his hot temper arose. He ordered the family to get off the place in fifteen minutes. They were on the run, guiltily, in less than five. Although there was no evidence of a plot, Clay immediately suspected that the treachery of the White family was instigated by his political opponents.

Though relieved to learn the truth about poor Launey's

illness, Clay had more trouble ahead. Perry White, angered at the summary dismissal of his parents, publicly vowed to "kill Cash Clay." But the Lion of White Hall, who could not fight Fate, knew how to handle rash young men. He armed himself with a pistol and his sharp bowie knife and kept his watchdogs alert. When Perry attempted to make good his threat, he was no match for the old man. On Sunday morning, September 30, 1877, Clay and his son mounted a mule to ride into the village. Not far from the mansion Clay saw the Negro hiding in a pasture. Quickly dismounting, he drew his pistol and ordered White to make no move while he questioned his presence upon the estate. Suddenly the young man ran toward Clay and drew a weapon. At his first move Cassius fired twice and hit him with both shots. Perry White fell dead. Leaving the body with the blood caking upon the gaping wounds, Clay went to Foxtown and surrendered himself. When his trial came up, the jury acquitted him by defining the deed as justifiable homicide in self-defense. The Perry White affair, which Clay described as further evidence of a plot against him, became another chapter in the growing legend of the Lion of White Hall.[9]

It was not only in violent self-defense that Cassius Clay earned his reputation. In politics, although he repeatedly protested that he had retired, he continued to roar. "If I am the old lion they say I am," he bellowed, "I will show them that I have not lost my teeth or my claws." In January, 1877, he served as presiding officer of the Democratic state convention at Louisville, where he tried to attract a following upon the issue of corruption in the civil service. "In my sincere conviction the Democratic party is the only party which can reform the civil administration," he declared, pounding the table in his best campaign style. But reform—a recurrent issue in postwar American politics—got him nowhere.[10]

In 1880, although Kentucky Democrats did not choose

him as a delegate to the national convention, an honor he earnestly sought, he would not remain upon the sidelines. He campaigned in Kentucky for the Democratic nominees Hancock and English, as much from habit, it seemed, as from conviction. Many heard the "gray-haired old veteran," and Cassius reveled in the publicity he received. In his speeches he indicated that he had not changed his viewpoint in forty years of stumping the state. "I rejoice that the South is solid," he said, but he mourned that the section lagged behind the industrial North. Southerners should invite men of skill and capital into their states, he advised, and he urged the manufacture of cotton textiles. "The nation that makes and manufactures cotton will be the nation of the world . . . we must buy as much as possible from home manufactures. . . ." Even to the end of a long career, Clay remained a loyal advocate of a southern industry.[11]

Clay remained faithful to his Whig principles, but could not make up his mind which wing of the rejuvenated Whig Party he liked better. Four years after lauding the determination of the South to remain "solid" as the best guarantee against its becoming an economic colony of the North, he returned to the Republican Party, and denounced the South. "I come back to the Republican party with all the principles of my life intact," he said in 1884. Clay's difficulty was that both the "Bourbon" Democrats of the South and the northern Republicans professed allegiance to the Hamiltonian economic principles of the prewar Whig Party. Campaigning in New England for the "Plumed Knight," James G. Blaine, Clay renounced his years of "wandering" among the opposition. "I have been imploring the Democrats for these eighteen years to become civilized and to abandon their barbarous ideas . . . ," he said. He explained that he left the party "because I have not the right of equality in the Democratic party. . . ."

Clay boasted to northern voters that he had "tried to make the Democratic party better, but it was a hard task. . . ." So hard, in fact, that he gave it up and returned to the Republicans. The Democrats snubbed him, he complained; even worse, they consistently lost elections. So Clay packed up his baggage and moved back into his old organization. With an old man's pride he pointed out that 1884 marked the eighth canvass in which he had addressed the citizens of the North. In every election in which he had participated, he said, the issues had been the same. "I stand here today in conflict with that enormous and disastrous power that has been our foe since 1854, then the slave power, now the solid South." In four years, Clay had completely reversed himself. His insatiable ambition would not let him rest: it had driven him from playing a leading role in the creation of the solid South to fighting it from the stump.[12]

But in spite of his feverish convolutions and his energetic campaigning, he was once more on the losing side. Clay's intuition failed with age, but he was not too old to find an excuse for Blaine's defeat. "The Union flag went down in disaster," he mourned, and he blamed the Radical policy of coercion against the South. "The attempt to Republicanize the rebel States by this means proved a dead failure," he said. Thus Cassius Clay explained his final disappointment, and with that he prepared for the end of his life. Though he kept up a sporadic correspondence with political acquaintances, he took no more active part in politics.[13]

Clay's first activity after concluding the 1884 campaign was the writing of his autobiography. He was worried for fear his part in the events of the nineteenth century would go unnoticed by posterity. He first attempted to hire someone to write his biography, but when he could not, he laboriously produced his memoirs in the first person. To justify the work, he explained, "I desire to stand before the reader, and receive such consideration among men as my share in their triumph shall merit." On the title page he put as a motto the Latin phrase, "Quorum, pars fui"—"Of them, I was a part." To prove his important role in the antislavery crusade and the war which followed, he carefully examined his scrapbooks of clippings and his letter-files for information. In doing so he destroyed all non-complimentary or questionable materials. He was determined to present an unblemished—even angelic—appearance to posterity.[14]

The *Memoirs* which he published in 1886 maintained that determination. Clay wrote in ponderous style, with the long sentences and involved clauses so dear to his era. His point of view was entirely subjective: he saw himself as the innocent victim of designing men who took credit for his contributions, while he was a selfless patriot combatting wicked enemies. He devoted much space in the volume to refute the numerous "calumnies" which his own erratic antics caused. It was clear that Clay intended the autobiography to be the final vindication of his career.[15]

Clay presented his career, as he looked back over seventy years of his life, in the terms he thought most acceptable to the postwar intellectual climate. In 1885, when he was explaining why he had embarked upon a career of opposition to slavery fifty years earlier, he was writing for an audience more apt to appreciate an emotional religious denunciation of slavery than a logical economic argument against it. Clay conveniently ignored the reasons which had led him to conclude that the slave system was inimical to the interests of his state; he adopted, in his *Memoirs,* the romanticized story that he had been "converted" to abolitionism in a New Haven church by William Lloyd Garrison himself. Such an explanation was out of character, for Clay always deeply despised the religious antislavery movement. He was not, in an orthodox sense, religious, and when he portrayed himself as merely a follower of Garrison, he did not do justice to his own understanding of the weaknesses of slavery. In depicting the economic considerations which motivated Clay to attack slavery the *Writings,* published in 1848, are more satisfactory than the *Memoirs* of 1886.

While Clay's friends congratulated him upon the publication of the *Memoirs,* he had fewer and fewer regular visitors as the years passed. People had other interests; White Hall was isolated, off the main road; and the old man found it increasingly difficult to travel. He busied himself by writing articles for agricultural journals on such subjects as bee-keeping and the genealogy of Kentucky bluegrass. He became an amateur historian and a leading figure in the Kentucky State Historical Society, which he had helped to found. He read a paper on "Money" before the Filson Club, a historical association in Louisville, and he wrote articles for the newspapers.

In 1890, he addressed the sixteenth annual reunion of the Ohio State Association of Mexican War Veterans, where he proclaimed his expansionism: he advocated the political union of North and South American countries into one nation. He also wrote petitions to the Kentucky legislature, praying them to resist the encroachments of the corporation, which he declared was more dangerous to republican institutions than primogeniture. He told the assemblymen that the "Rail-Road power is too strong for a Republic, and that one or the other must die!" It was the same vehemence he had demonstrated against chattel slavery; only the incubus had changed.[16]

Despite his earnest efforts to remain in public life, Clay became more lonely. Launey was away now, and the aging man found fewer human contacts. The occasional visitor to White Hall met a sturdy old gentleman with shaggy hair, and came away with memories of the lovely mirrored reception room, Clay's pet flowers, the oleander, and the taste of his host's fine wine. But it was apparent that Clay was becoming childish and eccentric. So few friends were left to Clay that he associated with tobacco-growing tenant families who lived along the nearby Kentucky River. One such person was Dora Richardson, a fifteen-year-old orphan, whom Clay took into his home as a serving girl. Dora captured the imagination of the eighty-four-year-old general, and after a few months' courtship he married her. When a newspaper reporter asked to see his child bride, Clay responded that he would have to clean her up before he would expose her to the public. There was a tradition that Dora had never worn shoes prior to her wedding. Whether true or not, the story illustrated the girl's peasant background.[17]

Clay's marriage to a minor aroused the self-appointed protectors of public morals. A band of them, led by the sheriff of Madison County, called at White Hall to rescue Dora from what they regarded as an immoral situation. Gallantly Cassius fought for his bride, as he had for Mary Jane—under quite different circumstances—so many years before. He fortified his "thirty room armed castle" against the invasion. The two brass cannons which had remained unused in the *True American* office in 1845 now stood upon the White Hall lawn, their menacing snouts guarding the mansion. The Lion still had claws, and he bared them to the posse. Testily he told his unwanted visitors that he had legally married Dora. Moreover, he said, as a Kentucky gentleman he had never kept a woman against her will, and he would not do so now. If the girl wished to leave, she was free to do so; but if she chose to stay—his dulling eyes flashed with the fire of old as he lit a match and held it over a cannon—then he would protect her. From a second-story window Dora shouted that she would not leave White Hall and her aged husband. Hearing that, as tradition has it, Cassius punctuated his command that the intruders leave by firing one of the cannons at them. They promptly decamped and left the bride and groom to continue their honeymoon in privacy.[18]

After a short period of "cleaning up" and instruction in etiquette, Dora accompanied her husband upon his journeys about the country. Unmindful of the gossip he had aroused by his unusual wedding—there was a difference of almost seventy years between their ages—Cassius proudly introduced Dora as the girl he "loved better than any woman he ever saw." But despite his amorous attentions, the second marriage was soon a failure. In July, 1897, after nearly three years with him, Dora moved out of White Hall. A year later Clay sued her for divorce on grounds of separation, and he wrote into his deposition that he had "fully met and discharged all the covenants of his marriage contract. . . ." If Dora was dissatisfied with her octogenarian husband, Cassius wanted the court to know that it was not because his manhood was failing. In August, 1898, he received his second divorce and immediately began the search for a third wife.[19]

Clay did not even wait for his divorce judgment before he renewed his quest. Time was short, he feared; he would waste none of it in an ordinary search. In March, 1898, in a letter to the press, he publicly proposed matrimony, and he received more than a thousand offers. "I seek a companion," he confessed, "and prefer one over 40 years old, but all ages allowed." His plea was widely circulated, and he received more replies than he could answer. It was a ribald, lecherous interest from sensation-hunters, however, which Clay now aroused, and not the fame of a bold political philosopher.[20]

Clay's advertisement for a companion was to no avail, as none of the offers resulted in marriage. He hired an eighteen-year-old boy as his bodyguard and companion and contented himself with watching Dora from a distance. He gave her a house in Pinkard, a village in Woodford County, and she married Riley Brock, who was much nearer her own age. But Clay paid no attention to "Brock and Company," as he called Dora's second husband. A railroad ran through Clay's estate to Pinkard, and the train operators regularly stopped at White Hall to take aboard the elderly, white-bearded gentleman. Each time he went, he would take some memento from the house to give her—a piece of silverware, a vase, or a lamp.

Despite his generosity, he was unable to entice Dora back to him, but he never stopped trying. Two weeks before he died he heard that Dora's second husband had died, and he invited her to come live with him for the rest of his life, but she refused.[21]

Cassius tried to win Dora back, for after she left him there were few people whom he trusted. As he passed his eighty-fifth birthday, his mind became more and more unstable. He became obsessed with the idea that someone was trying to kill him, and he redoubled his guard over the house, which he now called "Fort General Green Clay," after his father. Many a well-meaning friend would turn into his gate, a half-mile from the house, only to see Clay step from the porch with his rifle and shout, "Go back to town." One such visitor, who had come all the way from Cincinnati, after seeing the evidence of Clay's sharp eyes and hearing his booming voice, said, "Well, I didn't get to see the Lion of White Hall, but I've heard him roar." Others coming to call would find the house shuttered and locked, and no amount of pounding would cause the old man to open the door.

Such tales, and they were legion, added to the Clay legend and increased the apprehensions of those whose business took them to White Hall. One young man, delivering a telegram, saw Clay under an elm tree near the house. When the boy approached, the old man whipped out a pistol and grimly ordered the youth to "Halt! Take your hands out of your pockets!" But when he learned the boy's business, Clay pressed two drinks upon him before he would let him leave. There were other stories of Clay's violence, now made more fearsome by his uncertain and perhaps insane suspicions. A widely circulated account, in varying versions, concerned a certain Biggerstaff, who challenged the Lion to a duel. After the affair was arranged, however, Clay backed out. "If I go down there and kill him, it would be just a dead damn rascal,"

White Hall in 1894. Taken November 13, 1894, by Lexington photographer Isaac C. Jenks. From an original negative in the collection of J. Winston Coleman, Jr.

he airily expained, "but if he kills me, it would be a good man gone." [22]

Clay's mental lapses made him suspicious even of his own household servants. Caretakers and cooks would not long remain because of his abusive, threatening attitude. Upon one occasion his son, Brutus, went to investigate his condition and found that he had driven away his servants—a white couple who had been with him for twenty years—and had lived for three days on salt and eggs. The son quickly arranged for new servants to care for him, but it was not long before they too aroused Clay's suspicions. Since the son had located them, he imagined that his own children were members of a "vendetta" which plotted to kill him. Fearing poison, he distrusted food which had been prepared for him. He became more of a recluse and drove away his relatives when they came to visit him, summarily ordering even his children off the estate.[23]

Clay considered that everyone was his enemy, but he did not intend to go down without a fight. He would not leave his fort without weapons ready, and he wore a bowie knife constantly upon his person and concealed another in his bed. Reflecting that the state owed him protection which it did not provide, the old man refused to pay his taxes. When H. H. Collyer, the sheriff, obtained a judgment against Clay for unpaid property tax, he found it exceedingly difficult to execute. He first sent a deputy to take the document to the Lion of White Hall. Clay saw the man approaching and without asking any questions began firing at him. The deputy hid behind a tree while Cassius repeatedly hit it with his bullets. When he escaped, the sheriff mobilized a posse. Clay let the group get close enough to talk to him, but he would not leave the shelter of his porch. The sheriff shouted that he was required by law to deliver the judgment for the unpaid taxes. "I get no protection," the old man boomed in reply, "why should I pay taxes?" Without stopping to debate the matter the sheriff threw the document in Clay's direction, and the posse retreated ingloriously while Clay fired his cannon at them. High Sheriff Josiah F. Simmons' report to the county judge about the incident leaves no doubt of the vigor with which Clay defended his fortress:

Dear Judge: I am reportin about the posse like you said I had to. Judge we went out to White Hall but didn't do no good. It was a mistake to go out there with only 7 men. Judge, the old General was awful mad. He got to cussin and shootin and we had to shoot back. The old General sure did object to being arrested. Don't let nobody tell you he didn't and we had to shoot. I thought we hit him 2 or 3 times, but don't guess we did. He didn't act like it.

We come out right good, considerin. I'm having some misery from two splinters of wood in my side. Dick Collier was hurt a little when his shirt-tail and britches was shot off by a piece of horse shoe and nails that came out of that old cannon. Have you see Jack? He wrenched his neck and shoulder when his horse throwed him as we were getting away.

Judge, I think you will have to go to Frankfort and see Gov. Brown. If he would send Capt. Longmire up here with 2 light fielders he could divide his men—send some with the cannon around to the front of the house not too close, and the others around through the corn field and up around the cabins and spring house to the back porch. I think this might do it.[24]

His eccentric behavior made it possible for the truculent Clay to live at White Hall in the privacy he craved, but it also enabled his family, anxious to invalidate the will over which he had labored for years, to get a court order declaring him insane.

Finally the day came when the old general could no longer arise from his bed In the summer of 1903, when he was nearly ninety-three, he became ill. His kidneys were diseased, and he suffered from urinary sepsis because of an enlarged prostate. His physician, a woman, decided to send to Lexington for other doctors. Two physicians, Tom Bullock and Waller Bullock, journeyed to White Hall in a buggy. When they arrived they found the house tightly closed and were unable to arouse anyone. Aware of the old man's penchant for firing at intruders, Tom Bullock said, "Hell, let's get out of here," and they fled. In a few days, however, they received word that they had not been kept out intentionally, and they were invited to return. This time they were admitted through the back door and Cassius, in bed, apologized for the former rebuff. They diagnosed the disease, saw that his physician was following the recommended treatment, and left.[25]

The news quickly circulated through the Bluegrass that "old Cash" was dying. At the same time, the papers described the lingering illness of Pope Leo XIII in the Vatican. Sporting Kentuckians placed their bets on the "death Derby," and it gave Clay a childish satisfaction to know that he had outlived the pope. Even his illness did not calm his fighting temperament. On one of the last days of his life, with the heat of a July afternoon beating down upon the house, Clay became irritated at a large fly buzzing around the molding on the ceiling of the library where he lay. He ordered his servant boy to bring his rifle. Propping himself up in bed and taking careful aim, Cassius fired. His eye was as sure as it had ever been; the fly was splattered over the ceiling. To his death, Cassius Clay displayed an unyielding resistance and a good aim.[26]

A few days after he had silenced the noisy fly in a characteristic blaze of anger, the gaunt frame was still. At 9:30 P.M. on July 22, 1903, Cassius Marcellus Clay died. At last his long fight was over.

A few hours later, in the early morning hours of July 23, an electrical storm swept over central Kentucky, and a bolt of lightning knocked the head off the lofty statue of Henry Clay in the Lexington cemetery. Kentuckians, nodding their heads in awe, declared that "old Cash" had done it. He always was a fighter, people would say. The legend was now complete, and the Lion of White Hall passed into Kentucky folklore.[27]

REFERENCE MATTER

NOTES

Sources discussed in the Note on Sources are referred to by short titles in the Notes; all others are given in full at the first citation in each chapter.

CHAPTER I

1 Clay, *Memoirs*, I, 21. Clay intended to publish a second volume containing his writings and speeches, but it never appeared.
2 See *The Country Gentleman* (Albany, N.Y.), July 16, 1857, July 1, 1858, and comment by editor of the Louisville *Bulletin* (Ky.), February 24, 1884.
3 James G. Wilson and John Fiske (eds.), *Appleton's Cyclopaedia of American Biography* (6 vols., New York, 1888), I, 639–40. To "clear out" land in pioneer Kentucky involved making a survey, establishing some kind of settlement upon it, and clearing the patent records. For an example, see the deposition of Samuel Estill in the litigation, *Nelson's Heirs* vs. *Estill's Heirs*, 1810, in the private collection of Cassius M. Clay, Paris, Ky.
4 See William Chenault, "Early History of Madison County," in *Register of the Kentucky State Historical Society*, XXX (1932), 129.
5 Clay, *Memoirs*, I, 39; C. Frank Dunn, "General Green Clay in Fayette County Records," in *Register of the Ky. State Hist. Soc.*, XLIV (1946), 146; "Certificate Book," *ibid.*, XXI (1923), 6; *Biographical Encyclopaedia of Kentucky of the Dead and Living Men of the Nineteenth Century* (Cincinnati, 1878), p. 353; "Excerpts from the Journal of Governor Isaac Shelby," in *Register of the Ky. State Hist. Soc.*, XXVIII (1930), 19.
6 Clay, *The Clay Family*, p. 87; Clay, *Memoirs*, I, 20; William H. Perrin (ed.), *History of Bourbon, Scott, Harrison, and Nicholas Counties, Kentucky* (Chicago, 1882), p. 454. Green and Sally Clay's children were as follows: Elizabeth Lewis, who married

J. Speed Smith; Paulina, who married William Rodes; Sally Ann, whose second husband was Madison C. Johnson; Sidney Payne, Brutus Junius, and Cassius Marcellus. Another daughter, Sophia, died in infancy.—Clay, *The Clay Family*, p. 90.

7 Clay, *Memoirs*, I, 17. Another explanation for the unusual names was that "there had been too many Johns, Charleses, and Henrys in the family."—Cassius M. Clay, Paris, Ky., in a conversation with the author, summer, 1951, quoting a family tradition which stemmed from Green Clay.

8 Clay, *Memoirs*, I, 29–30.

9 *Ibid.*, 21, 45; 22–23, 31–34.

10 *Ibid.*, 24.

11 *Ibid.*, 35, 362–63; Cassius M. Clay to Brutus J. Clay, February 10, 1828, in Brutus J. Clay Papers.

12 Cassius to Brutus Clay, Feb. 10, 1828; Clay, *Memoirs*, I, 37.

13 *Kentucky Reporter* (Lexington), Nov. 19, 1828; Will of Green Clay is in Will Book D, pp. 461–67, Madison County, Ky. The will is dated July 18, 1828, and there are codicils dated July 22, August 14, and September 3, 1828. *Biographical Encyclopaedia of Kentucky*, p. 353; Clay, *Memoirs*, I, 37.

14 *Ibid.* 39–40; Green Clay to Sidney Payne Clay, September 2, 1818, in Sidney P. Clay Papers, The Filson Club, Louisville, Ky.

15 Clay, *Memoirs*, I, 23; Sally Clay to Cassius Clay, August 6, 1849, in the Cassius M. Clay Collection, Lincoln Memorial University. Mrs. Clay remarried soon after Green Clay's death, and lived for nearly forty years thereafter, until her ninetieth year. Clay, *Memoirs*, I, 19; G. Glenn Clift, *Kentucky Marriages, 1797–1831* (Frankfort, 1939), p. 325, quoting *Kentucky Reporter*, April 20, 1831.

16 Ulrich B. Phillips, *Life and Labor in the Old South* (Boston, 1929), p. 79; Coleman, *Slavery Times in Kentucky*, pp. 81, 120, 156; Clay, *Memoirs*, I, 46.

17 *Kentucky Reporter*, March 31, 1830; Record Book containing minutes of the Board of Trustees of Transylvania University, in Transylvania University Library, pp. 70–71 (October 6, 1828), and entry for September 15, 1827; Cassius M. Clay letter, March 9, 1898, in *The Transylvanian*, XV (1907), 452–54; *ibid.*, I (1829), 360; Alva Woods, *Intellectual and Moral Culture, A Discourse Delivered at His Inauguration as President of Transylvania University* (Lexington, 1828), p. 23. For a general history of Transylvania University, see Robert and Johanna Peter, *A*

History of Transylvania University (Louisville, 1899), and Alvin F. Lewis, *Higher Education in Kentucky* (Washington, 1899). Also helpful are the Minutes of the Union Philosophical Society, in Transylvania Library, and Clay, *Memoirs*, I, 35, 46–47.

18 Librarian's Check-Out Journal, Transylvania University Library, 1818–1834, entries for January 7 and January 28, 1829; *The Transylvanian*, I (1829), 360.

19 Clay, *Memoirs*, I, 46–47, 55.

20 Cassius M. Clay letter, March 9, 1898, previously cited; *Argus of Western America* (Frankfort, Ky.), May 13, 1829; *Kentucky Reporter*, May 13 and May 20, 1829; *The Transylvanian*, I (1829), 200; "Address to the General Assembly of the Commonwealth of Kentucky," in Record Book containing minutes of the Board of Trustees of Transylvania University, December 11, 1829. Although he served on a committee which solicited funds to restore the losses of the Union Philosophical Society, Clay kept his secret well. For an account of the university's religious difficulties, see Robert and Johanna Peter, *A History of Transylvania University*.

21 Clay, *Memoirs*, I, 47, 62–64.

22 Cassius M. Clay to Brutus J. Clay, June 19, 1831, March 27, 1831, both in Brutus J. Clay Papers.

23 Clay, *Memoirs*, I, 47–48, 52–53.

24 *Ibid.*, 47, 55, 59, 61.

25 The author is grateful to an unnamed researcher in the Yale University Library for information concerning academic life in New Haven while Clay was a student there, in a letter to the author, November 21, 1951. See also Clay, *Memoirs*, I, 54–55.

26 Clay to Brutus J. Clay, December 4, 1831, in Brutus J. Clay Papers. Clay's hearty appetite received its share of attention in New Haven, although he found eating habits strange. He reported to his brother that he ate "oysters and cod fish plenty, but little more yet. . . . I board with a good old sort of woman, who always asks me to take something of everything at the table—there is a certain dish of crackers poked at me after I have finished the dessert, much to my grief! I can bear much. When after having asked me to take everything on the table and at last asked me to have my milk a 'little warmed' I blubbered out!"—*Ibid.*

27 Clay to Brutus J. Clay, June 19, 1831, in Brutus J. Clay Papers.

28 Clay, *Memoirs*, I, 55–57.

29 Clay, *Writings*, pp. 174–75.

30 Clay to Brutus J. Clay, December 4, 1831, in Brutus J. Clay Papers.
31 The speech is printed in Clay, *Writings*, pp. 38–43. Quoted passages are from pp. 40 and 41. See also Clay, *Memoirs*, I, 57.
32 Clay, *Writings*, p. 43.

CHAPTER II

1 *Speech of Robert Wickliffe, Delivered before a Mass Meeting of the Democracy of Kentucky, at the White Sulphur Spring, in the County of Scott, on September 2, 1843* (Lexington, 1843), p. 15. Earlier, Robert Wickliffe, Sr., had been a Henry Clay Whig. See *Lexington Intelligencer*, January 23, 1838.
2 Cassius and Henry Clay were second cousins; their paternal grandfathers were brothers. See Clay, *The Clay Family*, for a family genealogy.
3 Clay, *Memoirs*, I, 73.
4 *Ibid.*, 71–72. Rollins later moved to Missouri, where he became a Unionist Congressman during the Civil War. He and Cassius Clay maintained a close friendship.
5 Clay, *Memoirs*, I, 71–72; "John P. Declary to the Public," February 25, 1833, and "Cassius M. Clay to the Public," March 8, 1833, both in *Observer and Reporter* (Lexington, Ky.), March 14, 1833. Declary responded with a rebuttal, dated March 18, which appeared in the issue of March 28, 1833.
6 Marriage Book Number 1, Fayette County Court House, Lexington, p. 115; G. Glenn Clift, *Kentucky Marriages, 1831–1849* (Frankfort, 1939), p. 23.
7 Clay, *Memoirs*, I, 72–73.
8 *Lexington Intelligencer*, April 8, 1834; *Observer and Reporter*, April 17, May 29, 1834; "Cassius M. Clay to the Citizens of Madison County," May 19, 1834, in *Lexington Intelligencer*, May 23, 1834. The minimum age requirement for members of the General Assembly was the subject of much contention in Kentucky. Cassius interpreted the constitution to mean that if he had attained his twenty-fourth birthday before the legislature convened he would meet the age requirement. His critics, however, declared that to be eligible he must have passed his twenty-fifth birthday prior to the session.
9 *Lexington Intelligencer*, March 21, 1834, March 21, and June 9, 1835; *Observer and Reporter*, June 11, 1834, April 8 and 15, 1835.

Some of Clay's associates in the Richmond and Lexington Turnpike Road Company who later denounced him were Robert Wickliffe, Sr., Waller Bullock, who participated in the expulsion of Clay's press in 1845, and Squire Turner, whose son Cyrus was a victim of Clay's bowie knife in a brawl in 1851. See the *Lexington Intelligencer*, March 21, 1835, for a complete listing of the road company commissioners.

10 *Lexington Intelligencer*, March 21, 1834; *Observer and Reporter*, June 11, 1834. Whigs resented President Andrew Jackson's strictures on borrowed capital. "And shall one universal ruin overtake all the noble enterprises of the day," one of them asked, "just to gratify the ante-diluvian, absurd ideas of an ignorant and prejudiced old man?"—*Observer and Reporter*, September 28, 1836. For further information on the southern Whig Party, see Arthur C. Cole, *The Whig Party in the South* (Washington, 1913); Thomas D. Clark, *A History of Kentucky* (New York, 1937), pp. 218–22; Samuel M. Wilson, *History of Kentucky* (2 vols., Chicago and Louisville, 1928), II, 185–90; Ulrich B. Phillips, "The Southern Whigs, 1834–1854," in *Essays in American History Dedicated to Frederick Jackson Turner* (New York, 1910), p. 209; and Charles G. Sellers, Jr., "Who Were the Southern Whigs?" in *American Historical Review*, LIX (1954), 335–46.

11 Clay, *Memoirs*, I, 73; *Lexington Intelligencer*, August 7, 1835; *Journal of the House of Representatives of the Commonwealth of Kentucky, 1835–1836* (Frankfort, 1836), pp. 4, 25–27, 29–31, 94, 176–77, 236, and 243.

12 Cassius M. Clay, "Speech, in the House of Representatives of Kentucky, upon a bill to take the sense of the people of this Commonwealth, as to the propriety of calling a Convention," in Clay, *Writings*, pp. 45–46.

13 Clay, *Memoirs*, I, 73–74.

14 Clay to Brutus J. Clay, August 5, 1836, in Brutus J. Clay Papers; *The Ohio Farmer* (Cleveland), April 30, 1859, p. 143; *The Country Gentleman* (Albany, N.Y.), May 31, 1860, carries an advertisement of Clay's: "C. M. Clay, Breeder of Pure Short-Horn Cattle, Southdown sheep, and Essex and Spanish Pigs." In 1860, Clay lectured upon the five major breeds of cattle and also upon "Breeding as an Art," at the Yale Agricultural Lectures, which attracted nation-wide attention among farmers and stock-raisers. See *The Country Gentleman*, February 2, 1860, p. 31.

15 Clay to Brutus J. Clay, December 19, 1837, in Brutus J. Clay Pa-

pers; *Journal of the House . . . 1837–38*, pp. 45, 83, 157, and 242; Robert Wickliffe, *An Address to the People of Kentucky on the subject of the Charleston and Ohio Railroad* (Lexington, 1838), p. 19; Clay, "Speech on the Bill conferring Banking Privileges on the Charleston, Cincinnati, and Louisville Railroad Company, before the committee of the Whole, in the House of Representatives of the Commonwealth of Kentucky," in Clay, *Writings*. Quoted passages are from pp. 54–55. See also *Lexington Intelligencer*, January 12, 1838, for an account of the speech.

16 Clay, *Memoirs*, I, 78; *Journal of the House . . .*, pp. 68–69, 162; *Lexington Intelligencer*, January 9, February 6, 1838. On January 6, 1838, Sprigg objected to the report of the banking committee, of which Clay was a member. "The gentleman from Shelby made a forcible speech . . . which often met with applause from all parts of the House," a correspondent reported. "He gave the Committee some hard knocks." It may have been those verbal blows which Cassius returned. See *Lexington Intelligencer*, January 9, 1838.

17 *Ibid.*

18 *The True American* (Lexington, Ky.), July 22, 1845, in Clay, *Writings*, p. 278.

CHAPTER III

1 *Observer and Reporter* (Lexington, Ky.), June 24, 1840; Clay, *Memoirs*, I, 74; Speech of B. Mills Crenshaw of Barren County, before the Kentucky Legislature, in *Frankfort Commonwealth*, January 26, 1841.

2 *Speech of R. Wickliffe on the Negro Law, August 10, 1840* (Lexington, 1840), pp. 11–13, 21.

3 *Observer and Reporter*, July 8, 11, and 15, 1840.

4 *Ibid.*, July 15, 1840.

5 *Speech of R. Wickliffe*, p. 5.

6 Clay to *Lexington Intelligencer*, in Clay, *Writings*, p. 134.

7 C. M. Clay, *A Review of the Late Canvass . . .* (Lexington, 1840), pp. 4–6.

8 *Ibid.*, p. 14.

9 *Speech of R. Wickliffe*, pp. 7 and 14.

10 Clay, *Memoirs*, I, 74; Clay, *Review of the Late Canvass . . .*, p. 4; *Speech of R. Wickliffe*, pp. 13–14; *Speech of R. Wickliffe in reply to the Rev. R. J. Breckinridge . . . on the 9th November,*

1840 (Lexington, 1840), pp. 4 and 45; Clay to Brutus J. Clay, August 4, 1840, Brutus J. Clay Papers.

11 Clay, *Writings*, pp. 71–74.

12 *Observer and Reporter*, July 18, 1840.

13 Clay to Brutus J. Clay, April 19, 1841, Brutus J. Clay Papers. Cassius wrote an almost identical note the following day, indicating his concern over the Wickliffe attack.

14 John C. Breckinridge to Robert J. Breckinridge, May 18, 1841, and Laeticia Breckinridge to Mrs. R. J. Breckinridge, May 11, 1841, both in Breckinridge Family Papers; Elisha Warfield to Brutus J. Clay, April 27, 1841, Brutus J. Clay Papers.

15 E. Warfield to Brutus J. Clay, April 27, 1841, and J. Speed Smith to Brutus J. Clay, April 29, 1841, both in Brutus J. Clay Papers; Redd to Robert J. Breckinridge, May 5, 1841, Breckinridge Family Papers. Clay later said that the duel occurred because Wickliffe introduced Mrs. Clay's name into the debate, to which he took exception as "inadmissable."—Clay, *Memoirs*, I, 80–81.

16 *Western Citizen* (Paris, Ky.), May 21, 1841; Agatha Marshall to Robert J. Breckinridge, May 18, 1841, Breckinridge Family Papers; Clay, *Memoirs*, I, 81.

17 Z. F. Smith, "Duelling, and some noted duels by Kentuckians," in *Register of the Kentucky State Historical Society*, VIII (1910), 77–87. Quoted passage is from pp. 79–80.

18 Clay, *Writings*, pp. 317–18. Clay's oratorical addition was not always consistent. Sometimes there were 500,000, and sometimes 600,000, non-slaveholders in Kentucky.

CHAPTER IV

1 Clay, *Writings*, pp. 86–87, 90.

2 "Letters to the *Lexington Intelligencer*, written during the pendency, before the Senate of Kentucky, of a bill from the House of Representatives, Repealing the Laws of 1833, 1840, and 1794, Prohibiting the Slave Trade" (1843), in Clay, *Writings*, p. 118.

3 Clay to editor of the New York *World*, February 19, 1861, photostat in possession of the author. The letter was marked "Confidential." *The True American*, February 11, 1846. "Toast" copied from the *Trumbull Democrat*, in *The True American*, January 14, 1846.

4 Clay to Salmon P. Chase, December 21, 1842, in Chase Papers, Historical Society of Pennsylvania.

5 Clipping from the *New York Tribune*, in File Box #2, Cassius M. Clay Collection, Lincoln Memorial University.

6 *Cincinnati Daily Gazette*, August 7, 1843; S. M. Brown to Clay, October 20, 1843, in *Observer and Reporter* (Lexington, Ky.), April 3, 1850. Delphton is now named Donerail.

7 Two accounts of the affair, one by Clay and one by Brown, appeared in the *Cincinnati Daily Gazette*, August 7 and August 19, 1843. Because of Brown's hasty charge and his use of the pistol, Clay always felt that the outsider had been brought in especially to kill him. Brown, on the other hand, accused Clay of needlessly maiming him. See *Observer and Reporter*, April 3, 1850, and Clay, *Memoirs*, I, 82–85.

8 The eye-witness report of the fight is worth quoting in full: "You have heard of the tremendous fight at the Cave. It was not slow. It was the first Bowie knife fight I ever saw, and the way my cousin *Cash* used it was tremendious [*sic*]. Blows on the head hard enough to cleave a man's skull asunder, but Brown must have a skull of extraordinary thickness. He stood the blows as well if not better than the most of men would do. Cassius most gallantly faced and even advanced on his [Brown's] six barrel revolving pistol, which alone saved his life. He sprang in upon him and used the knife with such power that Brown was either paralyzed by the blows, or forgot his revolver. I parted them, but have declined giving a written statement about it."—Thomas A. Russell to William H. Russell, August 27, 1843, in Special Collections, University of Kentucky Library.

9 *Cincinnati Daily Gazette*, October 18, 1843; Clay to S. M. Brown, October 10, 1843, in *Observer and Reporter*, April 3, 1850. The suit was *Commonwealth of Kentucky* vs. *Cassius M. Clay*, Order Book 29, p. 300, September 30, 1843, cited in Coleman, *Slavery Times in Kentucky*, p. 306; Clay, *Memoirs*, I, 86–89. Quoted passage is from p. 89.

10 *Cincinnati Daily Gazette*, March 19, 1844 and November 4, 1843.

11 Clay to Salmon P. Chase, January 19, 1844, Chase Papers, Hist. Soc. of Pa.

CHAPTER V

1 Millard Fillmore to Thurlow Weed, and Washington Hunt to Weed. in *Thurlow Weed Memoirs* (2 vols., Boston, 1884), II,

112–13; Lewis Tappan to Clay, July 6, 1844, Tappan Letter Book #4, p. 59, Tappan Papers, Library of Congress; Henry B. Stanton to Chase, February 6, 1844, in *Salmon P. Chase Diary and Correspondence* (Washington, 1903), pp. 462–65; Clay to Brutus J. Clay, January 6, 1844, Brutus J. Clay Papers. Original copies of the "free papers" which Clay obtained for his emancipated slaves are in the collection of J. Winston Coleman, Jr., Lexington, Ky.

2 "Speech against the annexation of Texas to the United States, delivered at Lexington, Kentucky, on the 13th day of May, 1844, in reply to Thomas F. Marshall," in Clay, *Writings*, pp. 97–116. Quoted passages are from pp. 97, 111, 113, and 116. The speech was first printed in *Observer and Reporter* (Lexington, Ky.), June 1, 1844.

3 Fillmore and Hunt to Weed, in *Thurlow Weed Memoirs*, II, 112–13.

4 Clay to W. I. McKinney, mayor of Dayton, Ohio, March 20, 1844, in *Cincinnati Daily Gazette*, April 4, 1844. Clay also offered other reasons for supporting his cousin: "Were I an Ohioan I might perhaps go with you, but as a Kentuckian I am for Clay, a bank, a binding link between the states, a tariff, and the division of the proceeds of the public lands among the states for the purposes of paying state debts and educational needs or for congress taking them to pay the same debts upon some more speedy plan and redeeming us in the eyes of the world from the damning doctrine of . . . repudiation."

5 "Emancipation: Its Effect," in *Letters of Cassius M. Clay* (New York, 1843); Clay, *Writings*, p. 60; *Cincinnati Daily Gazette*, April 4, 1844; May 9, 1845; *Cassius M. Clay and Gerrit Smith: A Letter of Cassius M. Clay . . . to the Mayor of Dayton, Ohio, with a Review of It by Gerrit Smith of Peterboro, N.Y.* [Utica, 1844].

6 Clay to Colonel J. J. Speed of Ithaca, N.Y., July 10, 1844, in Clay, *Writings*, pp. 158–59.

7 *Observer and Reporter*, May 22, 25, 29, and July 10, 1844.

8 *Ibid.*, July 10, 1844.

9 Clay, *Memoirs*, I, 92–93.

10 In 1837, Green Clay, their first child to survive his infancy, was born; in 1839 came Cassius Marcellus, Jr., the second son to bear his father's name. The daughters were named after their grandmothers: Mary Barr and Sarah Lewis. Another daughter, Flora, died in infancy. See Clay, *The Clay Family*, for Cassius and

Mary Jane Clay's children and grandchildren. Other information cited is in *The True American*, June 3, 1845, and *Observer and Reporter*, October 18, 1845. The entrusted slaves continued their malevolence against his family. Two years later, Emily stood trial for the attempted poisoning of Clay's new-born infant son, but was acquitted of the charge. *Commonwealth of Kentucky* vs. *Emily, a slave*, Fayette County Circuit Court, File 1103, April 15, 1845. For a full account of the Emily case see Coleman, *Slavery Times in Kentucky*, pp. 264–66.

11 Conversation with Miss Helen Bennett of Richmond, Ky., a granddaughter of Cassius Clay, summer, 1950. See also Clay to Brutus J. Clay, March 19 and April 3, 1843, Brutus J. Clay Papers, for examples of Clay's paternal concern for his children.

12 Mrs. Mary Jane Clay to Mrs. Llewellyn P. Tarleton, August 28, 1844, written from Cleveland, in Cassius M. Clay Collection, The Filson Club.

13 Henry Clay to editor, *Observer and Reporter*, September 2, 1844, in the issue of September 4; Henry Clay to Cassius Clay, September 18, 1844, in Clay, *Memoirs*, I, 101–2; Henry Clay to Joshua Giddings, September 21, 1844, in Giddings-Julian Papers. In repudiating Cassius' statements, Henry set forth his views favoring state control of slavery, even in the District of Columbia.

14 Clay, *Memoirs*, I, 100–101.

15 E. D. Mansfield, *Personal Memories* (Cincinnati, 1879), p. 223; Clay to editor, *Cincinnati Daily Enquirer*, April 8, 1853, in the paper dated April 9.

16 Speech of Cassius M. Clay at Cleveland, in *Cincinnati Daily Gazette*, September 7, 1844.

17 *Cincinnati Daily Gazette*, September 7, 1844; Theodore Foster to Birney, September 12, 1844, in D. L. Dumond (ed.), *Letters of James G. Birney* (2 vols., New York, 1938), II, 842.

18 "Speech, at Tremont Temple, September 19, 1844, after the adjournment of the great convention on Boston Common," in Clay, *Writings*, p. 160. The speech also appeared in the *Cincinnati Daily Gazette*, October 3, 1844. Earlier in the day, Clay took part in the gathering on Boston Common and delivered a "short" speech. He appeared there with Daniel Webster and Georgia Senator John M. Berrien.

19 Clay, *Memoirs*, I, 99–100; Washington Hunt to Weed, in *Thurlow Weed Memoirs*, II, 123.

20 Clay, "Address to the People of Kentucky, January, 1845," in

Clay, *Writings*, p. 173; "Appeal to Kentucky and to the World," in *The True American*, October 7, 1845, and copied into Clay, *Writings*, p. 312. Lewis Tappan to James G. Chester, October 1, 1844, Tappan Letter Book #4, p. 155, Tappan Papers, Library of Congress; Theodore Foster to Birney, September 12, 1844, in Dumond (ed.), *Birney Letters*, II, 842; *Anti-Slavery Bugle* (Salem, Ohio), October 31, 1845.

21 *Cincinnati Daily Gazette*, October 31, 1844; Clay, *Memoirs*, I, 101–3.

22 Clay to editor, *Boston Atlas*, in *Cincinnati Daily Gazette*, January 24, 1845. Clay said that Henry Clay took his defeat with "ill grace," and at a dinner in Lexington, Henry severely criticized ex-Governor (then Senator) James T. Morehead and the abolitionists of New York for his failure. "I could but feel that part of his censure was against myself," Cassius admitted. See Clay, *Memoirs*, I, 103–4.

CHAPTER VI

1 Clay, *Writings*, p. 181.

2 *Ibid.*, p. 182.

3 Clay to Thomas B. Stevenson, editor of the *Frankfort Commonwealth*, January 8, 1845, in *Observer and Reporter* (Lexington, Ky.), January 15, 1845.

4 Prospectus for *The True American*, in *Observer and Reporter*, February 19, 1845. It was reprinted in Clay, *Writings*, pp. 211–12.

5 The fortification of *The True American* office is described in Clay, *Memoirs*, I, 107. Further information is in *A Pioneer Emancipator* (Richmond, Ky., 1881), a pamphlet reprinted from the *Nashville American* (Tenn.). The critic quoted is Thomas F. Marshall, in the oration delivered to the mass meeting which suppressed Clay's paper, in W. L. Barre (ed.), *Speeches and Writings of Hon. Thomas F. Marshall* (Cincinnati, 1858), p. 202.

6 *The True American*, June 3, 1845, in Clay, *Writings*, pp. 213 and 217.

7 *Liberty Hall and Cincinnati Gazette*, June 10, 1845; *New York Tribune*, quoted in *Observer and Reporter*, July 16, 1845; Lewis Tappan to Clay, July 17, 1845, in Tappan Letter Book #4, pp. 359 and 363, Tappan Papers, Library of Congress.

8 Clay to Salmon P. Chase, July 3, 1845, Chase Papers, Historical

Society of Pennsylvania; *Observer and Reporter*, June 7, 1845; *Kentucky Compiler*, quoted in the *Liberator* (Boston), August 29, 1845. The complete anonymous communication, delivered through the courthouse, was as follows: "You are meaner than the autocrats of hell. You may think you can awe and curse the people of Kentucky to your infamous course. You will find when it is too late for life, the people are no cowards. Eternal hatred is locked up in the bosoms of braver men, your betters, for you. The hemp is ready for your neck. Your life cannot be spared. Plenty thirst for your blood—are determined to have it. It is unknown to you or your friends, if you have any, and in a way you little dream of. Revengers."

9 Clay, *Memoirs*, I, 106; *The True American*, June 10, 1845, in Clay, *Writings*, pp. 224–25.

10 *The True American*, June 10, July 15, 1845.

11 *The True American*, June 17 and July 1, 1845, in Clay, *Writings*, pp. 234 and 255.

12 *The True American*, June 17 and July 8, 1845, in Clay, *Writings*, pp. 238–40 and 266.

13 *The True American*, June 3 and 24, 1845; the latter material copied into Clay, *Writings*, p. 250.

14 Clay, *Writings*, p. 302; Martin, *The Anti-Slavery Movement in Kentucky Prior to 1850*, p. 115. The paper's circulation is the subject of some debate, as subscription lists do not seem to have survived. Clay's private report on his circulation is worth repeating: "I commenced with about 300 subs. in Ky. and have now about 600, a very encouraging increase, in as much as I have no agent in this State, trusting to the friends of the movement to come up slow but *sure*. If by the end of the year we have 1000 in the state we will have done great things and fully come up to our hopes—for one has to make great sacrifices in taking our paper, as he is cut off in *business* forthwith by the slaveocracy who have the *wealth*."—Clay to Salmon P. Chase, July 3, 1845, Chase Papers, Hist. Soc. of Pa.

15 Clay, *Writings*, p. 302; *The True American*, July 15, 1845, *ibid.*, p. 274.

16 *The True American*, July 15, 1845, in Clay, *Writings*, p. 274.

CHAPTER VII

1 *The True American,* August 12, 1845. Extracts of the article appeared in many newspapers, including the *Liberator* (Boston), September 5, 1845. Its author was Nathaniel Ware. Henry Wilson to Clay, November 6, 1871, in Cassius M. Clay Collection, Lincoln Memorial University.
2 *The True American,* August 12, 1845, reprinted in many other papers, including the *Observer and Reporter* (Lexington, Ky.), August 20, 1845, and the *Liberty Hall and Cincinnati Gazette,* August 28. Clay's fiery lines were also quoted in Thomas F. Marshall's oration on August 18; see W. L. Barre (ed.), *Writings and Speeches of Hon. Thomas F. Marshall* (Cincinnati, 1858), pp. 196–210.
3 From the *Louisville Journal,* in *Liberty Hall and Cincinnati Gazette,* August 21, 1845. See also Clay, *Writings,* pp. 303–4.
4 *Ibid.,* p. 305.
5 *Observer and Reporter,* August 20, 1845.
6 Clay, *Memoirs,* I, 108–9; Clay, *Writings,* p. 307; *The True American* Extra, August 15, 1845, copy in Lexington Public Library. It is also published in Clay, *Writings,* pp. 287–92.
7 *Ibid.,* pp. 292–94.
8 *Ibid.,* pp. 294–98; *Liberty Hall and Cincinnati Gazette,* August 28, 1845. In 1837, Elijah P. Lovejoy died from a gunshot wound at Alton, Illinois, as he directed the defense of his antislavery newspaper, the *Observer,* against the attack.
9 Clay, *Writings,* pp. 298–300.
10 The speech is in Barre, *Marshall,* 196–210. It also appeared in the *Observer and Reporter,* August 20, 1845, and was reprinted in many other papers. For the "official" account of the affair of August 18, see the anonymous pamphlet, *History and Record of the Proceedings of the People of Lexington and its Vicinity in the Suppression of The True American,* etc. (Lexington, 1845). For Marshall's activity in 1840, see *Observer and Reporter,* September 13, 1845. For a brief sketch of Marshall's life, see Barre, *Marshall,* pp. 7–12.
11 Barre, *Marshall,* pp. 201–7; *Observer and Reporter,* August 20, 1845.
12 *Observer and Reporter,* August 20, 1845; *Liberty Hall and Cincinnati Gazette,* August 28, 1845.

13 *Observer and Reporter*, August 20, 1845, in *Charleston Courier* (S.C.), August 28, 1845.

14 *New York Tribune*, August 25 and 26, 1845; *Cincinnati Gazette*, October 16, 1845; *Liberator*, September 19, 1845; Address, "Cassius M. Clay," by Judge Byron Paine, manuscript in Paine Papers, Wisconsin Historical Society.

15 *Anti-Slavery Bugle* (New Lisbon, Ohio), August 29, 1845; *Liberator*, August 29, 1845; Tappan to Gamaliel Bailey, Jr., September 1, 1845, Tappan Letter Book #4, p. 398, Tappan Papers, Library of Congress.

16 D. M. Craig to Rev. E. Berkley, August 19, 1845, in *Filson Club History Quarterly*, XXII (July, 1948), 197–98; Clay, *Writings*, p. 258 n. For a contemporary view of the suppression of Clay's paper, see *Filson Club History Quarterly*, XXIX (October, 1955), 320–23. Accounts of the incident are in Lowell Harrison, "Cassius Marcellus Clay and *The True American*," in the same journal, XXII (January, 1948); Clement Eaton, *Freedom of Thought in the Old South* (Durham, 1940); Coleman, *Slavery Times in Kentucky;* and William H. Townsend, *Lincoln and The Bluegrass* (Lexington, 1955).

CHAPTER VIII

1 *The True American*, November 4 and December 23, 1845, in Clay, *Writings*, pp. 336 and 368; Clay to Mr. Hartshorne, Sept. 5, 1845, in *Anti-Slavery Bugle* (Salem, Ohio), October 3, 1845.

2 *Observer and Reporter* (Lexington Ky.), October 4, 8, 1845. After returning from the Mexican War, Clay renewed his legal action against the committee and in a change of venue to neighboring Jessamine County received a judgment of $2500 against the group. "This gives the press to the defendants, and gives the plaintiff its value," the *Observer and Reporter* editor remarked, April 5, 1848.

3 *Ibid.*, September 27, 1845.

4 *Ibid.*, October 11, 1845.

5 *The True American*, April 8, 1846. As Clay explained it, "I never intended from the beginning to edit the paper and it is a sore task to me; I could do much more 'on the stump' and shall be glad to be relieved at some near future time of the painful *duty*."

—Clay to Chase, June 30, 1846, Chase Papers, Historical Society of Pennsylvania.

6 Broadside, "To the Laborers of Kentucky," copy in Brutus J. Clay Papers; also printed in *The True American*, March 11, 1846.

7 Clay to Chase, July 3, 1845, Chase Papers, Hist. Soc. of Pa.; *Cassius M. Clay's Appeal to all the Followers of Christ in the American Union* (Lexington, December 9, 1845). There is a copy in the library of the Philosophical and Historical Society of Ohio in Cincinnati. *The True American*, March 4, 1846; and May 13, 1846, in Clay, *Writings*, p. 454; issues for December 23, 1845, January 6, February 4 and 18, 1846. For other religious pronouncements against slavery, see the issues for March 25, April 1 and 22, and May 13, 1846.

8 *The True American*, February 4, March 11, and April 1, 1846; and March 4, 1846, in Clay, *Writings*, p. 398.

9 *The True American*, March 4, 11, 1846; Clay's review of the *Autobiography of John G. Fee*, dated August 24, 1896, is in Clay Scrapbook #2, in the private collection of Professor J. T. Dorris, retired, of Eastern Kentucky State College, Richmond, Ky.

10 Clay to Louis Marshall, November 20, 1895, in Berea College Library, Berea, Ky.

11 The remainder of the prayer deserves repetition: "Do thou, O Lord, tighten the chains of our black brethren, and cause slavery to increase and multiply throughout the world. And whereas many nations of the earth have loved their neighbors as themselves, and have done unto others as they would that others should do unto them, and have broken every bond, and have let the oppressed go free, do thou, O God, turn their hearts from their evil ways, and let them seize once more upon the weak and defenceless, and subject them to eternal servitude! And O God! although thou has commanded us not to muzzle the poor ox that treadeth out the corn, yet let them labor unceasingly without reward, and let their own husbands, and wives, and children, be sold into distant lands without crime, that thy name may be glorified, and that unbelievers may be confounded, and forced to confess that indeed thou art a God of justice and mercy! Stop, stop, O God, the escape from the prison house, by which thousands of these '*accursed*' men flee into foreign countries, where nothing but tyranny reigns; and compel them to enjoy the unequalled blessings of our own *free* land!

"Whereas our rulers in the Alabama legislature have emancipated a black man, because of some eminent public service, thus bringing thy holy name into shame, do thou, O God, change their hearts, melt them into mercy, and into obedience to thy will, and cause them speedily to restore the chain to that unfortunate soul! And O God, thou searcher of all hearts, since many of thine own professed followers—when they come to lie down on the bed of death, and enter upon that bourne whence no traveller returns, where every one shall be called to account for the deeds done in the body, whether they be good or whether they be evil—emancipate their fellow men, failing in the faith, and given over to hardness of heart, and blindness of perception to the truth, do thou, O God, be merciful to them, and the poor recipients of their deceitful philanthropy, *and let the chain enter in the flesh, and the iron into the soul forever!*"—*The True American*, March 25, 1846, in Clay, *Writings*, pp. 409–10.

12 *The True American*, November 4, 1845, in Clay, *Writings*, p. 336; issue for April 8, 1846, *ibid.*, p. 430; "Address, before the Board of Home Missions of the M. E. Church, at Musical Fund Hall, Philadelphia, January 14, 1846," in Clay, *Writings*, pp. 523–35; "Speech in Broadway Tabernacle, New York, January, 1846," in Clay, *Writings*, pp. 185–201.

13 *Ibid.*, pp. 200–201.

14 In 1894, a correspondent told Clay, "When as a boy I sat in the Broadway Tabernacle, New York, and heard you thunder on 'The rights of man,' I had no idea that I would ever be honored with a letter from you. . . . Your name has always been in my mind a synonym for heroism, patriotism, and eloquence."—T. Dewitt Talmadge to Clay, January 27, 1894, in Cassius M. Clay Collection, Lincoln Memorial University. The obituary was in Louisville *Times*, July 25, 1903.

15 *Observer and Reporter*, January 14 and February 11, 1846.

CHAPTER IX

1 James D. Richardson (ed.), *A Compilation of the Messages and Papers of the Presidents* (10 vols., Washington, 1896–99), IV, 442. For the beginning of the war with Mexico, see Justin H. Smith, *The War with Mexico* (2 vols., New York, 1919), I, 149–55, and Robert S. Henry, *The Story of the Mexican War* (Indianapolis and New York, 1950), pp. 47–52.

2 *The True American*, May 25, 27, 1846, in Clay, *Writings*, p. 470.

3 *Anti-Slavery Bugle* (Salem, Ohio), June 26, 1846; *The True American*, June 17, 1846.

4 *Anti-Slavery Bugle*, June 26, July 3, and August 21, 1846; *Christian World*, in *The True American*, July 8, 1846. See also Silas M. Holmes to Birney, March 27, 1847, in D. L. Dumond (ed.), *Letters of James G. Birney* (2 vols., New York, 1938), II, 1042.

5 Cassius M. Clay to *Cincinnati Herald*, June 28, 1846, in *Anti-Slavery Bugle*, July 17, 1846; Cassius M. Clay to *New York Tribune*, December 10, 1846, in Clay, *Writings*, pp. 477–78. *The True American*, July 8, 1846. For other evidence that Clay regarded his enlistment as more than a patriotic duty, see his speech, May 20, 1846, in Clay, *Writings*, pp. 475–76; Cassius M. Clay to *Christian Reflector*, *ibid.*, 483–87; and the *Autobiography of John G. Fee*, (Chicago, 1891), p. 127. See also Clay, *Memoirs*, I, 110, and Clay to Salmon P. Chase, June 30, 1846, in Chase Papers, Historical Society of Pennsylvania.

6 Clay to Chase, December 27, 1847, Chase Papers, Hist. Soc. of Pa.; *The True American*, July 8, 1846; Memorandum of Cassius Clay's business, May 24, 1846, in Brutus J. Clay Papers. *The True American* did not long survive Clay's absence. Its subscription list declined so that it no longer paid expenses. Clay's lawyer therefore ordered it stopped. *Observer and Reporter* (Lexington, Ky.), October 24, 1846; *The True American*, October 21, 1846 (final issue).

7 *The True American*, July 8, 1846; Clay, *Memoirs*, I, 117–18; "Speech at Lexington, May 20, 1846," in Clay, *Writings*, p. 476.

8 *Observer and Reporter*, July 25, 1846; Clay to Brutus J. Clay, June 22, 1846, Brutus J. Clay Papers; Clay, *Memoirs*, I, 118. Jackson became a lieutenant in the company.

9 *Observer and Reporter*, June 3, 6, 1846.

10 Clay, *Memoirs*, I, 540; Clay to Brutus J. Clay, June 22, 1846; Mary Jane Clay to Brutus J. Clay, October 11, 1846, both in Brutus J. Clay Papers. A son, Brutus Junius, was born to Mary Jane Clay on February 20, 1847.

11 *Observer and Reporter*, June 13, November 4, 1846.

12 *Ibid.*, June 27, 1846. Humphrey Marshall (1812–72) graduated from West Point in the class of 1832 and resigned his commission the following year. After service in the Mexican War he was sometime congressman from Kentucky, minister to China, general in the Confederate Army, and a member of the Confederate

Congress. A brief sketch of his life is in James G. Wilson and John Fiske (eds.), *Appleton's Cyclopaedia of American Biography* (6 vols., New York, 1888), IV, 226–27.

13 *Louisville Courier*, in *Observer and Reporter*, September 27, 1848; Clay, *Memoirs*, I, 119. Although he had successfully completed his mission, Clay's difficulties with the Madame were only beginning. He had entered the house under orders to remove the deserters, and the woman had resisted his mission, but when the war was over she sued him for damages to her property. Cassius promptly branded the suit another attempt by his political enemies to discredit him. In the litigation, his counsel declared that he had acted under military orders and that every person involved in the affair was as guilty as he. Clay's plea did not affect the jury, however, who held him personally responsible for the damage and awarded the woman a settlement of $501 against him. He appealed the matter to Congress and eventually received a full reimbursement.

14 *Observer and Reporter*, July 8, 18, 1846; Cassius M. Clay to Brutus J. Clay, June 22, 1846, in Brutus J. Clay Papers.

15 *Observer and Reporter*, July 8, 18, October 14, November 11, 1846; letter of Lieutenant J. S. Jackson, September 23, *ibid.*, November 7, 1846. Lieutenant Jackson's vehement complaints elicited a response from the Army Paymaster Department. See *ibid.*, November 4, 1846.

16 *Ibid.*, October 14, November 7, 1846.

17 *Ibid.*; Clay, *Memoirs*, I, 119–34.

18 *Register of the Kentucky State Historical Society*, XXIV (1926), 197; Clay, *Memoirs*, I, 140–41. Clay declared that at the threat of a duel with him, Marshall had tried to drown himself but was fished out. A short time later, Captain Marshall duelled with Lieutenant James S. Jackson, who had surrendered his commission in order that Cassius could have a command. Jackson wounded Marshall and then resigned his commission and returned to Kentucky. Marshall remained, at the risk of a court-martial for participating in a duel. *Observer and Reporter*, October 31, 1846.

19 *Ibid.*, November 7, 1846; Clay, *Memoirs*, I, 135.

20 Letter, September 21, in *Observer and Reporter*, October 14, 1846; *ibid.*, October 31, 1846.

21 Letter, September 26, *ibid.*, November 7, 1846.

22 *Ibid.*, January 16, 20, and February 13, 1847.

23 Clay, *Memoirs*, I, 142–44; *Observer and Reporter*, March 13, 24,

1847. For Clay's defense of the surrender, see his letter to the New Orleans *Picayune*, July 15, in *Observer and Reporter*, August 21, 1847. See also Clay, *Writings*, pp. 480–82.

24 Letter of John P. Gaines, February 10, in *Observer and Reporter*, March 24, 1847; Clay, *Memoirs*, I, 145–47.

25 *Observer and Reporter*, March 24, 1847; Clay, *Memoirs*, I, 146–48.

26 The exact words Clay uttered at that time have not been preserved, and there are several versions available. The one used here is the inscription of a sword presented to Cassius by the "Citizens of Madison and Fayette Counties, Kentucky," 1848, which is in the Berea College Library, Berea, Kentucky. Five of Clay's fellow prisoners issued a public statement praising Clay's action, and they quoted him as saying, "Kill the officers—spare the soldiers." And when a Mexican major ran to him and pointed a cocked pistol at his breast, Cassius "still exclaimed—'Kill me—kill the officers—but spare the men—they are innocent!' "—Letter of A. C. Bryan, W. D. Ratcliffe, Charles E. Mooney, John J. Finch, and Alfred Argabright, to editor, *Observer and Reporter*, October 20, in the issue of October 25, 1847. See also letter of Lieutenant W. J. Heady, one of the prisoners, from Mexico City, May 12, in *Observer and Reporter*, June 19, 1847. But Clay's foes circulated another version by Captain C. C. Danley of the Arkansas Cavalry, who was also a prisoner. Danley depicted Clay as shamefully begging for his life: "For the sake of the great Henry Clay, who opposed the war, and who opposed the annexation of Texas, and whose relation I am—spare me! For the sake of the great Whig Party, a member of which I am, and which opposed the war and the annexation of Texas, and which has been neutral in this war, but which will rise against Mexico to a man, if I am killed—spare me!" The political bias in Danley's testimony is clear. His statement, dated May 29, is in *Observer and Reporter*, June 14, 1848. See also Clay, *Memoirs*, I, 267–68.

27 *Ibid.*, 149–51; letter of Bryan *et al.* in *Observer and Reporter*, issue of October 25; *Observer and Reporter*, October 23, 1847.

28 *Ibid.*, May 19, 26, 29, June 9, 27, July 8; and letter from a prisoner, May 24, in issue for July 3, 1847; Cassius M. Clay to Brutus J. Clay, June 18, 1847, from Mexico City, in Brutus J. Clay Papers; Clay, *Memoirs*, I, 152–55.

29 Cassius M. Clay to Brutus J. Clay, June 18, 1847, in Brutus J. Clay Papers.

30 Clay, *Memoirs*, I, 155–60.
31 *Ibid.*, 159–63.
32 *Ibid.*, 159; *Observer and Reporter*, April 10, 28, June 2, 23, October 23, 1847. In General Taylor's official report of the battle he said that the "Kentucky cavalry, under Colonel Marshall, rendered good service dismounted, acting as light troops on our left, and afterwards, with a portion of the Arkansas regiment, in meeting and dispersing the column of cavalry at Buena Vista." From Agua Nueva, March 6, in *Observer and Reporter*, April 28, 1847.
33 Jeptha Dudley to Brutus J. Clay, December 9, 1847, Brutus J. Clay Papers; *Observer and Reporter*, August 18, November 27, 1847. For examples of the fetes in Clay's honor, see Louisville *Examiner*, December 18, 1847; *Observer and Reporter*, December 22, 25, 1847. See also Clay, *Memoirs*, I, 164–67, footnotes.

CHAPTER X

1 Clay, *Memoirs*, I, 169.
2 Clay to Henry Clay, April 13, 1848, in New York *Courier and Enquirer*, reprinted in *Observer and Reporter* (Lexington, Ky.), April 22, 1848; *Cincinnati Daily Gazette*, April 21, 1848.
3 Quotations are from Holman Hamilton, *Zachary Taylor: Soldier in the White House* (Indianapolis and New York, 1941–51), pp. 52–56 and 67–68. Clay to Louisville *Times*, April 28, 1848, in Clay Scrapbook #2, in the collection of Professor J. T. Dorris, Richmond, Ky.; Clay, *Memoirs*, I, 168–69, written many years after the event, exaggerates Clay's part in the Kentucky Taylor movement.
4 *Observer and Reporter*, April 22, 1848; Clay, *Memoirs*, I, 169–70.
5 Clay to Salmon P. Chase, July 14, 1848, in Chase Papers, Historical Society of Pennsylvania.
6 Richard French to Howell Cobb, September 10, 1848, in U. B. Phillips (ed.), *Correspondence of Robert Toombs, Alexander H. Stephens, and Howell Cobb* (Washington, 1911), p. 126; Robert J. Breckinridge to Samuel Steel, April 17, 1849, copy in Breckinridge Family Papers; Clay, *Memoirs*, I, 177 n.
7 William H. Townsend, *Lincoln and the Bluegrass* (Lexington, 1955), p. 163.
8 Clay, *Memoirs*, I, 175–76; Louisville *Examiner*, January 13, 1849,

quoting Clay to editor, December 25, 1848, in Clay, *Memoirs,* I, 491–92; *Observer and Reporter,* January 20, March 17, 31, April 7, 1849.

9 *Cincinnati Daily Gazette*, April 7, 1849; *Observer and Reporter*, April 18, 1849; Breckinridge to Samuel Steel, April 17, 1849, Breckinridge Family Papers.

10 *Observer and Reporter*, April 28, 1849; Hambleton Tapp, "Robert J. Breckinridge and the Year 1849," in *Filson Club History Quarterly*, XII (1938), 125–50, especially pp. 134–35.

11 *Observer and Reporter*, April 28, 1849; Tapp, "Breckinridge," in *Filson Club Hist. Quart.*, XII, 134.

12 *Observer and Reporter*, May 5, 1849.

13 *Ibid.*, May 9, 1849.

14 Clay, *Memoirs,* I, 175–78 n.

15 *Ibid.*, 177–83; Louisville *Examiner*, in *Anti-Slavery Bugle* (Salem, Ohio), May 4, 1849.

16 *Observer and Reporter*, July 11, 1849.

17 *Ibid.*, July 7, 11, 1849; Clay, *Memoirs,* I, 185–87; Clay to Brutus J. Clay, July 21, 1849, in Brutus J. Clay Papers; Sally Clay Dudley to Clay, August 6, 1849, in Cassius M. Clay Collection, Lincoln Memorial University.

18 Clay, *Memoirs,* I, 187; *Observer and Reporter*, August 15, 1849; Coleman, *Slavery Times in Kentucky*, pp. 316–17.

19 *Frankfort Commonwealth*, March 11, 1849; Clay, *Memoirs*, I, 176.

20 *Ibid.*, 212; J. Speed Smith to Brutus J. Clay, February 24, 1851, in Brutus J. Clay Papers; Clay to William H. Seward, April 24, 1851, in Seward Papers.

21 Clay was referring to Daniel Webster and to Daniel S. Dickinson, Democrat of New York.

22 Clay to a Committee of Free-Soilers in Maine, May 26, 1851, in *Frankfort Commonwealth*, July 15, 1851; Clay to Seward, April 24, 1851, in Seward Papers. Clay was ever-confident. "The newspapers affect to regard my letter to the Maine Free Soilers as revolutionary, although I have a thousand times stated my position to be constitutional opposition to slavery," Clay told Seward. "This cannot be avoided. They have the press, the money, the education, the intelligence—great odds—yet I will not yield."—Clay to Seward, August 8, 1851, in Seward Papers.

23 *Frankfort Commonwealth*, March 11, May 27, and July 1, 1851; Clay, *Memoirs*, I, 212; Clay to Seward, April 24, 1851, in Seward Papers.

24 *Kentucky Statesman* (Lexington), March 1, 1851; *Frankfort Commonwealth*, March 11, 1851. For other ex-emancipationists who rejected the Clay-Blakey ticket, see *Frankfort Commonwealth*, April 1, May 27, and July 22, 1851; for other Whig ridicule, see the issue of April 15, 1851.

25 John G. Fee to Gerrit Smith, 1851, in *Calendar of the Gerrit Smith Papers in the Syracuse University Library, General Correspondence, Vol. Two, 1846–1854* (2 vols., Albany, 1942), #602; *Frankfort Commonwealth*, May 6, 1851. Bailey expressed antislavery views similar to those long advocated by Cassius Clay. After years of agitation, Bailey thus summed up his attitude: "Workingmen of Kentucky, think of yourselves! See you not that the system of slavery enslaves all who labor for an honest living? You, white men, are the best slave property of the South, and it is your votes that makes [*sic*] you so."—Quoted in Coleman, *Slavery Times in Kentucky*, pp. 320–21.

26 *Anti-Slavery Bugle*, August 16, 1851. For a list of Clay's speaking engagements, see *Kentucky Statesman*, May 28 and July 9, 1851.

27 "Speech of C. M. Clay at Lexington, Ky., Delivered August 1, 1851." Pamphlet, n.d. Quoted passages are from pp. 2, 3, and 11. For examples of Blakey's campaign speeches, see clippings in George D. Blakey Scrapbook in the library of Western Kentucky State College, Bowling Green, graciously lent to the author by Mrs. Mary T. Moore, librarian.

28 *Frankfort Commonwealth*, September 2, September 16, 1851; Clay to Giddings, September 3, 1851, in Byron R. Long, "Joshua Reed Giddings, A Champion of Political Freedom," in *Ohio Archaeological and Historical Society Publications*, XXVII (Columbus, 1919), 33; Clay, *Memoirs*, I, 213. See also E. Merton Coulter, "The Downfall of the Whig Party in Kentucky," *Register of the Kentucky State Historical Society*, XXIII (1925), 164. The official returns of the election are in *Frankfort Commonwealth*, September 2, 1851.

29 Clay to Chase, August 12, 1851, Chase Papers, Hist. Soc. of Pa.

CHAPTER XI

1 Clay to Giddings, September 3, 1851, in Byron R. Long, "Joshua Reed Giddings, A Champion of Political Freedom," in *Ohio Archaeological and Historical Society Publications*, XXVIII (Co-

lumbus, 1919), 33; George W. Julian, *Political Recollections* (Chicago, 1884), p. 119; *Cincinnati Daily Gazette*, September 27, 1851; Clay to Seward, August 8, 1851, in Seward Papers. Clay to Chase, August 12 and August 27, 1851, in Chase Papers, Historical Society of Pennsylvania; Clay to Seward, February 17, 1849, in Seward Papers: "In you I hope for the leader of my faith and the perfecter of my ambition. . . ."

2 Richmond *Weekly Messenger* (Ky.), May 21, 1852, lists the Central Committee of the Kentucky Free-Soil Party as follows: W. P. Moore, Scion Kimball, J. H. Rawlings, C. M. Clay, Henry Hawkins, H. Doolin, G. C. Smith, Dr. J. Howard, R. Stapp, and Thomas Coyle. The same paper, June 11, 1852, lists the state party platform; Cassius M. Clay was its author. *Cincinnati Daily Gazette*, July 1, 1852, quoted a Washington correspondent who said, "Cassius M. Clay . . . will undoubtedly be the candidate for the Vice-Presidency." On the Julian tour in Kentucky, see John G. Fee to Julian, September, 1852, in Giddings-Julian Papers, for an itinerary; Julian, *Political Recollections*, pp. 125–27; and Clay, *Memoirs*, I, 500. For the results of the vote, see *Frankfort Commonwealth*, November 22, 1852. Cassius M. Clay to Committee planning an Anti-Slavery Convention in Cincinnati, March 25, 1853, in *Cincinnati Daily Gazette*, April 23, 1853.

3 Cassius M. Clay to Cincinnati committee, March 25, 1853, in *Cincinnati Daily Gazette*, April 23, 1853.

4 Clay, *Memoirs*, I, 230; *Cincinnati Daily Gazette*, March 16, 1854; Speech of Cassius M. Clay before the Young Men's Association of Chicago, July 4, 1854, a copy in the Cassius M. Clay Collection, The Filson Club; Clay to Seward, February 6, 1855, in Seward Papers; Clay to Chase, August 1, 1854, in Chase Papers, Hist. Soc. of Pa.

5 Cassius M. Clay to H. M. Knox, November 1, 1854, in Miscellaneous Personal Collection, Library of Congress; *Cincinnati Daily Gazette*, November 10, 1854; Clay to Seward, February 6, 1855, in Seward Papers.

6 Clay, *Memoirs*, I, 212; *Berea College, Kentucky, An Interesting History* (Cincinnati, 1883), pp. 6–12. In 1855, Fee's school became Berea College. When the minister admitted Negroes along with whites, Clay withdrew his support of the enterprise. See *Autobiography of John G. Fee* (Chicago, 1891), pp. 130–31, 138.

7 *Cincinnati Daily Gazette*, June 18, July 23, 1855; *Observer and Reporter* (Lexington, Ky.), July 18, August 22, 1855; *Frankfort*

Commonwealth, July 27, 1855; Clay to Samuel Evans, August 5, 1855, in Cassius M. Clay Collection, The Filson Club; Clay to Chase, June 4 and July 5, 1855, in Chase Papers, Hist. Soc. of Pa.

8 Clay to editor, *New York Tribune*, 1855; Clay to G. W. Brown, February 12, 1856, in the *Kansas Herald of Freedom;* both in the Cassius M. Clay Collection, The Filson Club. Clay to editor, *Cincinnati Daily Gazette*, July 19, 1855, in the issue dated July 23.

9 Clay to the Republican Association of Washington, February 8, 1856, in Cassius M. Clay Collection, The Filson Club; George W. Julian, "The First Republican National Convention," in *American Historical Review,* IV (1899), 319–20; Grace N. Taylor, "The Blair Family in the Civil War," in *Register of the Kentucky State Historical Society*, XXXVIII (1940), 286; *Proceedings of the First Three Republican Conventions of 1856, 1860, and 1864, Including Proceedings of the Antecedent National Convention Held at Pittsburg, in February, 1856, as Reported by Horace Greeley* (Minneapolis, 1893), pp. 9–11. Clay to Chase, May 10 and June 24, 1856, in Chase Papers, Hist. Soc. of Pa.

10 *Cincinnati Daily Gazette*, November 8, 9, 1854, February 13, 1855; C. C. Huntington, "A History of Banking and Currency in Ohio before the Civil War," in *Ohio Archaeological and Historical Society Publications*, XXIV (1915), 453–4; Clay, *Memoirs*, I, 537.

11 The deed of trust between Clay and his assignees, dated February 20, 1856, is in the Brutus J. Clay Papers; Clay to John R. Johnston, February, 1856, original in the collection of J. Winston Coleman, Jr., Lexington, Ky., and kindly lent to the author; *Cincinnati Daily Gazette*, March 1, May 10, 1856; *Observer and Reporter*, April 15, 1856; schedule of property sold at Cassius M. Clay's sale, in Brutus J. Clay Papers.

12 Clay to John R. Johnston, January 11, 1853, in the private collection of Foreman M. Lebold, Chicago, and a photostat in the collection of J. Winston Coleman, Jr., who kindly let the author copy it. Clay, *Memoirs*, I, 222–29.

13 *Ibid.*, 539–40; Clay, *The Clay Family;* Clay to John R. Johnston, February, 1856; Clay to Chase, May 10, 1856, in Chase Papers, Hist. Soc. of Pa.

14 Clay to *New York Tribune*, 1855, in Cassius M. Clay Collection, The Filson Club.

15 *Proceedings of the First Three Republican Conventions . . .*,

pp. 79–85; *Frankfort Commonwealth*, July 16, 1856; *Cincinnati Daily Gazette*, July 31, and August 11, 1856.

16 *Ibid.*, August 14, 19, 1856; Clay to Chase, October 9, 1856, from Elyria, Ohio, in Chase Papers, Hist. Soc. of Pa. For other accounts of Clay's speeches in the Midwest, see *Cincinnati Daily Gazette*, August 12, 13, 15, 21, and September 9, 1856. The issue of October 2 described Clay's address at the Louisville Courthouse. He also spoke at the Tippecanoe Battlefield, in Pittsburgh, and in Milwaukee; see issues for September 18 and October 4.

17 See the author's "Abraham Lincoln Deals with Cassius M. Clay," in *Lincoln Herald*, 55: No. 4 (Winter, 1953), 15–23.

18 *Speech of C. M. Clay before the Young Men's Republican Central Union of New York . . . , October 24th, 1856* [New York? 1856?], pp. 5–6.

19 *Ibid.*, p. 10. For other comments on Clay's Tabernacle speech, see Andrew W. Crandall, *The Early History of the Republican Party, 1854–1856*, pp. 70–71, and Ruhl J. Bartlett, *John C. Fremont and the Republican Party*, p. 48.

20 *Cincinnati Daily Gazette*, October 28, December 2, 1856; *Frankfort Commonwealth*, December 10, 1856.

21 Clay to Archibald W. Campbell, January 27, 1859, in *Calendar of the Francis H. Pierpont Letters and Papers in West Virginia Depositories* (West Virginia Historical Records Survey, 1940), Item #14, original in West Virginia University Library. *Frankfort Commonwealth*, August 10, 31, 1859; Clay to editor, Richmond *Messenger* (Ky.), December 28, 1859, clipping in scrapbook, "Political Issues, 1858–60," Vol. V, in the library of the Philosophical and Historical Society of Ohio, Cincinnati.

22 Clay to Chase, March 6, 1857, April 14, 1859, in Chase Papers, Hist. Soc. of Pa.; Clay to Seward, April 18, July 10, 1858, in Seward Papers.

23 *Speech of Cassius M. Clay, at Frankfort, Kentucky, from the Capitol Steps, January 10, 1860* (Cincinnati, 1860).

24 Frank W. Ballard to Clay, October 28, 1859, and Cephas Brainerd to Clay, December 19, 1859, in Cassius M. Clay Collection, Lincoln Memorial University; Charles C. Nott to Lincoln, February 9, 1860, in David C. Mearns (ed.), *The Lincoln Papers* (2 vols., New York, 1948), I, 229; *New York Times*, February 16, 1860.

25 Leslie Combs to John J. Crittenden, March 5, 1860, in John J. Crittenden Papers, Library of Congress.

26 Clay to Weed, March 8, 1860, in Weed Collection, University of Rochester.
27 *Proceedings of the First Three Republican Conventions . . .*, pp. 160–63; *Frankfort Commonwealth*, May 21, 1860.
28 Clay to Lincoln, May 21, 1860, in Mearns, *Lincoln Papers*, I, 246.

CHAPTER XII

1 Cassius M. Clay to Abraham Lincoln, May 21, 1860, in David C. Mearns (ed.), *The Lincoln Papers* (2 vols., New York, 1948), I, 246; Clay to W. Kenneau, editor of the *Cincinnati Commercial*, May 23, 1860, in Robert Todd Lincoln Collection; Clay to Salmon P. Chase, May 26, 1860, in Chase Papers, Library of Congress; Clay to Seward, May 21, 1860, in Seward Papers.
2 Perhaps the most perplexing problem in the interpretation of Cassius M. Clay's public career is his testimony that Lincoln promised him a cabinet post in 1860. Later, bitter at his failure, Clay repeatedly declared that the candidate had specifically promised him the office of Secretary of War should the party win. But at the time Clay made his allegations, Lincoln was dead and unable to refute or to corroborate Clay's statements, and there is no record of any such promise in Lincoln's files. In his *Memoirs* (I, 303) Clay said that Lincoln wrote him a letter, asking him to participate in the campaign, and promising him the office. Clay further declared that he deposited that letter in the Kentucky Historical Society as permanent proof of his argument, but the document has never been located in the archives of the society.

In his letter of January 10, 1861, to Lincoln, in the Robert Todd Lincoln Collection, Clay himself accepted the quoted words of Lincoln as the promise of a cabinet post, indicating that he considered the words as a definite commitment. The obvious inference is that Clay was so eager for the position that he read into the phrase more than the canny Lincoln intended.
3 *New York Times*, July 13, 1860; *Cincinnati Daily Gazette*, July 10, 11, 1860; John D. Defress to Clay, July 14, 1860, and Abraham Lincoln to Clay, July 20, 1860, both in Cassius M. Clay Collection, Lincoln Memorial University.
4 *Cincinnati Daily Gazette*, July 17, 19, 20, August 1, 1860; for Clay's Indiana schedule see *ibid.*, July 7, 18, 1860.

5 George W. Gans, George W. Waite to Lincoln, November 30, December 1, 1860; Curtis Knight to George D. Blakey, December 15, 1860, all in Robert Todd Lincoln Collection. See also James G. Birney to Clay, November 24, 1860, in Cassius M. Clay Collection, Lincoln Memorial University.

6 John G. Nicolay Memorandum, November 16, 1860, cited in Baringer, *A House Dividing: Lincoln as President-Elect*, pp. 52–53. Nicolay registered Breck as "a Judge Black of Ky."

7 George Robertson to John J. Crittenden, December 16, 1860, in *Life of John J. Crittenden, Edited by his Daughter, Mrs. Chapman Coleman* (2 vols., Philadelphia, 1871), II, 222; Cassius M. Clay to Lincoln, January 10, 1860, in Robert Todd Lincoln Collection; Clay, *Memoirs*, I, 215–16, 302–3.

8 Clay to the Republicans of Miami County, Ohio, November 26, 1860, in Cassius M. Clay Collection, The Filson Club; Clay to J. W. Gordon, December 19, 1860, in the New York *World*, clipping in Clay Scrapbook #2 in the private collection of Professor J. T. Dorris of Richmond, Ky., and also reprinted in *New York Times*, January 7, 1861.

9 Clay to Lincoln, January 10, 1861, in Robert Todd Lincoln Collection.

10 Rollins to Clay, January 19, 1861, in James M. Wood, Jr., "James Sidney Rollins: Civil War Congressman from Missouri" (Master's thesis, Stanford University, 1947), pp. 26–27.

11 *New York Times*, January 25, 28, February 2, 1861; Speech of Cassius M. Clay in Washington, January 26, 1861, copy in Cassius M. Clay Collection, The Filson Club; Chase to Clay, January 23, 1861, in Cassius M. Clay Collection, Lincoln Memorial University; Lincoln to William Kellogg, congressman from Illinois, and member of the Congressional Committee of Thirty-Three to consider compromise proposals, December 11, 1860, in John G. Nicolay and John Hay (eds.), *Abraham Lincoln: Complete Works . . .*, (2 vols., New York, 1920), I, 675–78; Clay to Lincoln, February 6, 1861, in Robert Todd Lincoln Collection.

12 *Ibid.*

13 William H. Seward to Lincoln, March 11, 1861, in Mearns, *Lincoln Papers*, II, 478; *New York Times*, March 13, 14, 15, 1861; Cassius M. Clay to Brutus J. Clay, March 13, 1861, in Brutus J. Clay Papers; telegram, Clay to Montgomery Blair, March 27, 1861, in Mearns, *Lincoln Papers*, II, 493; Clay, *Memoirs*, I, 255–57.

14 Clay to Lincoln, March 28, 1861, in Mearns, *Lincoln Papers,* II, 495.

15 *New York Times,* April 16, 19, 1861.

16 Tyler Dennett (ed.), *Lincoln and the Civil War in the Diaries and Letters of John Hay* (New York, 1939), p. 8; cited in Jay Monaghan, *Diplomat in Carpet Slippers* (Indianapolis and New York, 1945), p. 77.

17 *New York Times,* April 19, 20, 1861; Captain Lewis Towns to Clay, April 19, 1861, in Cassius M. Clay Collection, Lincoln Memorial University; Clay, *Memoirs,* I, 259–64, recounts several anecdotes about Clay's ferocious patriotism; J. W. Wright to Clay, April 20, 1861, and Clarence Eytinge to Clay, April 22, 1861, both in Cassius M. Clay Collection, Lincoln Memorial University; for a statement by one of the Clay Guards, see *Lincoln Lore,* #102 (March 23, 1931); Clay, *Memoirs,* I, 284.

18 Clay to Lincoln, May 11, 1861, in Mearns, *Lincoln Papers,* II, 605.; Clay to Lincoln, March 1, 1862, in Robert Todd Lincoln Collection.

19 Frank W. Ballard to Clay, May 11, 1861, in Cassius M. Clay Collection, Lincoln Memorial University; Sarah Agnes Wallace and Frances Elma Gillespie (eds.), *The Journal of Benjamin Moran* (2 vols., Chicago, 1948–49), I, 810.

20 Clay to the London *Times,* May 17, 1861, in *New York Times,* June 5, 1861.

21 *New York Times,* June 3, 21, 22, 1861; Francis Lieber to Clay, June 5, 1861; Elliot C. Cowdin to Clay, June 8, 1861; George W. Morgan, U.S. Minister to Portugal, to Clay, June 11, 1861; George F. Train to Clay, August 13, 1861, all in Cassius M. Clay Collection, Lincoln Memorial University.

22 *New York Times,* June 11, 13, 14, 18, 1861; Speech of Cassius M. Clay in Paris, May 29, 1861, copy in Cassius M. Clay Collection, The Filson Club; Worthington C. Ford (ed.), *Letters of Henry Adams, 1858–1891* (Boston, 1930), p. 92; cited in Monaghan, *Diplomat in Carpet Slippers,* p. 106.

23 Clay to Seward, June 7 and 21, 1861, in *Diplomatic Correspondence of the United States, 1861–1863* (3 vols., Washington, 1864), I, 286–89; Prince Gortchacov to Clay, January 8, 1862, in Cassius M. Clay Collection, Lincoln Memorial University; Monaghan, *Diplomat in Carpet Slippers,* p. 107.

24 Clay, *Memoirs,* I, 294–96.

25 This incident is fully described in Monaghan, *Diplomat in Carpet*

Slippers, p. 109; Bayard Taylor to Horace Greeley, July 5, 1862, cited in Harry J. Carman and Reinhard H. Luthin, *Lincoln and the Patronage* (New York, 1943), pp. 84–85.

26 William Cassius Goodloe to David S. Goodloe, July 19, 1861, cited in James Rood Robertson, *A Kentuckian at the Court of the Tsars: The Ministry of Cassius Marcellus Clay to Russia*, pp. 44–47.

27 Clay to Lincoln, July 25, 1861, in Robert Todd Lincoln Collection; Clay to Chase, November 22, 1861, in Chase Papers, Library of Congress; William L. Dayton to Clay, February 10, 1862, in Cassius M. Clay Collection, Lincoln Memorial University; Clay to Lincoln, March 1, 1862, in Robert Todd Lincoln Collection.

28 Clay to Lincoln, March 1, 1862, in Robert Todd Lincoln Collection.

29 Norman B. Judd to Clay, March 29, 1862, in Cassius M. Clay Collection, Lincoln Memorial University; Clay to Lincoln, June 17, 1862, in Robert Todd Lincoln Collection; Lincoln to Clay, August 12, 1862, in Clay, *Memoirs*, I, 304.

CHAPTER XIII

1 Speech of Cassius M. Clay in Washington, August 12, 1862, from the *New York Herald*, August 15, 1862, clipping in Cassius M. Clay Collection, The Filson Club; *New York Times*, August 13, August 14, 1862; Clay to Lincoln, August 13, 1862, in Robert Todd Lincoln Collection; Clay, *Memoirs*, I, 305–9. Much of the material in this chapter appeared in the author's "Abraham Lincoln Deals with Cassius M. Clay: Portrait of a Patient Politician," *Lincoln Herald*, 55: No. 4 (Winter, 1953), 15–23.

2 S. C. Pomeroy to Clay, August 13, 1862, and Wendell Phillips to Clay, August 19, 1862, both in Cassius M. Clay Collection, Lincoln Memorial University; Clay, *Memoirs*, I, 582–84.

3 *New York Times*, August 20, 1862; Clay to Stanton, August 13, 1862, in Edwin M. Stanton Papers.

4 Gideon Welles, *Diary of Gideon Welles* (3 vols., Boston and New York, 1911), I, 70; William B. Hesseltine, *Lincoln and the War Governors* (New York, 1948), pp. 249–57.

5 Theodore C. Pease, and James G. Randall (eds.), *The Diary of Orville Hickman Browning* (2 vols., Springfield, 1925 and 1933), I, 594–95, entry for December 12, 1862; Clay, *Memoirs*, I, 310.

6 Clay, *Memoirs,* I, 310–12; *New York Times,* August 26, 27, 1862; Clay to John A. Andrew, March 20, 1863, John A. Andrew Papers, complains that he was relieved of his command because he "refused to return fugitive slaves from my ranks."

7 Speech of Cassius M. Clay as Messenger of the President of the United States, in the Hall of the House of Representatives, at Frankfort, in the *Cincinnati Daily Gazette,* August 31, 1862, clipping in Cassius M. Clay Collection, The Filson Club; *New York Times,* September 6, 1862; Clay, *Memoirs,* I, 312.

8 *Ibid.* See also marginal notes by Clay in *Speech of Cassius M. Clay before the Law Department of Albany University, February 3, 1863,* (New York, 1863), pp. 13 and 24, copy in Cassius M. Clay Collection, Lincoln Memorial University; Clay to George Bancroft, October 28, 1866, in the Massachusetts Historical Society.

9 *New York Times,* September 25, 1862; speech of Cassius M. Clay in Washington, September 24, 1862, clipping in Cassius M. Clay Collection, The Filson Club; Clay, *Memoirs,* I, portrait facing p. 304.

10 Clay to Lincoln, September 29 and October 21, 1862, Robert Todd Lincoln Collection; also in Clay, *Memoirs,* I, 315; *New York Times,* October 15, 1862.

11 *Ibid.*, October 7, 8, 10, 11, 31, 1862; speech of Cassius M. Clay at Brooklyn, October 7, 1862, at the Brooklyn Academy of Music, clipping in the Cassius M. Clay Collection, The Filson Club.

12 Clay, *Memoirs,* I, 318–19; Clay to Lincoln, February 26, 1862, (two letters) in Robert Todd Lincoln Collection; *New York Times,* March 2, 9, 11, 12, 1863; Clay to William G. Smethers, March 15, 1863, in Huntington Library, San Marino, California; Clay to his wife, March 16, 1863, in Cassius M. Clay Collection, The Filson Club; Clay to John A. Andrew, Andrew Papers.

13 Clay to Chase, March 23, 1863, in Chase Papers, Library of Congress. Clay to Samuel Hallett and Co., April 8, 1863, in Robert Todd Lincoln Collection. The letter was a printed form.

CHAPTER XIV

1 Clay, *Memoirs,* I, 324.

2 Cassius M. Clay to Mary Jane Clay, April 30, 1863, in Cassius M. Clay Collection, The Filson Club; Clay, *Memoirs,* I, 354–56.

3 James R. Robertson, *A Kentuckian at the Court of the Tsars,* pp. 145–48. See also Clay to Salmon P. Chase, September 6, 1863, in the Chase Papers, Library of Congress.

4 Clay to Robert J. Walker, August 15, 1863, in Robert J. Walker Papers, Library of Congress.

5 Clay to Mary Jane Clay, July, 1863, in Cassius M. Clay Collection, The Filson Club; Frank A. Golder, "The Russian Fleet and the Civil War," in *American Historical Review,* XX (1915), 802–3; Robertson, *Kentuckian at Court of the Tsars,* pp. 148–67.

6 Clay to Henry Wilson, September 27, 1868, in Henry Wilson Papers; Clay, *Memoirs,* I, 246–48, 408–10, 465, 478. Clay to Brutus J. Clay, October 9, 1863, Clay Family Papers, Library of Congress. Clay's memory was faulty as to his position in 1860, when he still maintained an alliance with the New Yorker. After the nominating convention he wrote Seward, "You were my first choice and I so wrote to Chase's confidential friend, W. D. Gallagher . . . ; but it was difficult to get our delegation to vote for you . . . because Chase was in continual communication . . . with our friends. Yet he was my second choice. . . ."—Clay to Seward, May 21, 1860, in Seward Papers.

7 Clay to Lincoln, April 2, 1863, Robert Todd Lincoln Collection. Lincoln marked upon the envelope his summary of its contents: "Bitterly complaining of Mr. Seward's treatment of him, and especially with reference to money matters and the Secretary of Legation." See Clay to Seward, October 31, 1863, and October 17 and 22, 1864, cited in Joseph Schafer (ed.), *Memoirs of Jeremiah Curtin* ("Wisconsin Biography Series," Volume II [Madison, 1940]), p. 10; and Clay to Seward, March 30, 1863, in Seward Papers.

8 *Memoirs of Jeremiah Curtin,* pp. 31–78; Sarah Agnes Wallace and Frances Elma Gillespie (eds.), *The Journal of Benjamin Moran* (2 vols., Chicago, 1948–49), II, 1332.

9 *Memoirs of Jeremiah Curtin,* pp. 79–82.

10 *Ibid.,* pp. 175–76; Clay to Seward, May 19, 1863, in *Diplomatic Correspondence of the United States, 1861–1863* (3 vols., Washington, 1864), II, 791; Clay to Seward, June 17, 1863, *ibid.,* 795–96.

11 *Memoirs of Jeremiah Curtin,* pp. 176–77; E. D. Morgan to Hamilton Fish, November 28, 1871, in Hamilton Fish Papers.

12 Clay, *Memoirs,* I, 420, 538.

13 *Ibid.,* 363–406, 409.

14 Clay to Mary Jane Clay, December 26, 1863, in Cassius M. Clay Collection, The Filson Club; Count Orloff Davidoff to Clay, January 3, 1864 O.S., in Cassius M. Clay Collection, Lincoln Memorial University.

15 Clay, *Memoirs,* I, 435–37.

16 *Ibid.,* 346, 418–19; Count Adlerberg to Clay, July 28, August 9, 1864, Mme. Olga Barstchoff and Princess Nadine Galitzin to Clay, 1865, and B. Estvan to Clay, December 13, 1867, in Cassius M. Clay Collection, Lincoln Memorial University; Letter of Bayard Taylor, cited in Woldman, *Lincoln and the Russians,* p. 122.

17 Clay, *Memoirs,* I, 467.

18 Statement of Eliza Chautems, April 19, 1866, *ibid.,* 463–65.

19 Schuyler Colfax to Clay, May 18, 1867, *ibid.,* 471; Theodore C. Pease, and James G. Randall (eds.), *The Diary of Orville Hickman Browning* (2 vols., Springfield, 1925 and 1933) II, 141, entry for April 2, 1867.

20 Timothy Bombshell and others, *A Synopsis of Forty Chapters Upon Clay, not to be found in any treatise on the Free Soils of the United-States of America heretofore published* (n.d., n.p.), pp. 2, 5, 8, 10–11.

21 See the documents in Clay, *Memoirs,* I, 467–77.

22 *Ibid.,* 474.

23 Robertson, *Kentuckian at Court of the Tsars,* pp. 193–97; *Memoirs of Jeremiah Curtin,* pp. 103–28; Clay, *Memoirs,* I, 409–12.

24 *Memoirs of Jeremiah Curtin,* pp. 93–98; Clay, *Memoirs,* I, 414–16; Clay to Seward, February 6, 1866, cited in *Memoirs of Jeremiah Curtin,* pp. 15 and 99; Clay to Seward, April 23, 1866, in Seward Papers.

25 *Memoirs of Jeremiah Curtin,* p. 175; Clay to Henry Wilson, in Henry Wilson Papers; Clay to Seward, legation dispatch, January 30, 1868, cited in *Memoirs of Jeremiah Curtin,* p. 17; J. C. B Davis to Hamilton Fish, August 2, 1869, in Hamilton Fish Papers.

26 Clay to Schuyler Colfax, September 30, 1868, in Cassius M. Clay Collection, The Filson Club; Clay to Elihu B. Washburne, October 28, 1868, in Washburne Papers; Clay to Seward, April 26, 1867, in Seward Papers. See also Clay to Samuel Bowles, February 11, 1869, in *Filson Club History Quarterly,* XII, 3 (July, 1938), 167–69; Clay to Washburne, July 11, 1869, in Washburne Papers; and Clay to Washburne, July 21, 1869, in Massachusetts Historical Society.

27 *Memoirs of Jeremiah Curtin,* Editor's Introduction, pp. 16–23;

Clay to Seward, January 16, 1869, copy in Washburne Papers.

28 George Pomutz, American consul in St. Petersburg, to James R. Doolittle, November 6, 1867, cited in *Memoirs of Jeremiah Curtin*, p. 23; M. D. Landon to Clay, January 1, and January 16, 1868, both in Cassius M. Clay Collection, Lincoln Memorial University; Clay to John A. Andrew, February 18, 1867, in Andrew Papers.

29 Clay to Wilson, September 27, 1868, in Henry Wilson Papers; *Oration of Cassius Marcellus Clay before the Students and Historical Class of Berea College, Berea, Kentucky, October 16, 1895* (Richmond, Ky., 1895), pp. 7–10. Clay did not get his wish about the monument; it bears only his name.

30 Robertson, *Kentuckian at Court of the Tsars*, pp. 227–29; V. J. Farrar, *The Annexation of Russian America to the United States* (Washington, 1937), pp. 1–14.

31 Robertson, *Kentuckian at Court of the Tsars*, pp. 229–36; Woldman, *Lincoln and the Russians*, pp. 279–83; James M. Callahan, *The Alaska Purchase* (West Virginia University Studies in American History, Series I, Nos. 2 and 3 [Morgantown, 1908]), p. 21.

32 Clay to Samuel Bowles, February 11, 1869, in *Filson Club History Quarterly*, XII, 167–69; Timothy O. Howe to Hamilton Fish, December 15, 1871, in Hamilton Fish Papers.

CHAPTER XV

1 Clay to Chase, September 6, 1863, in Chase Papers, Library of Congress. Much of the material in this chapter appeared in the author's "Cassius M. Clay and the Cuban Charitable Aid Society," *Bulletin of the Historical and Philosophical Society of Ohio*, XII, No. 3 (July, 1954), 218–26.

2 Statement of Cassius M. Clay, October 18, 1880, in the Buffalo *Courier* (N.Y.), explaining his abandonment of the administration Republicans, clipping in Clay Scrapbook #2, p. 87, in the private collection of Professor J. T. Dorris, Richmond, Kentucky.

3 See William B. Hesseltine, "Economic Factors in the Abandonment of Reconstruction," *Mississippi Valley Historical Review*, XXII (1935), 191–210, for an interpretation of moderate opinion in the North.

4 Clay to Stanton, February 24, 1863, Edwin M. Stanton Papers; Clay to Chase, September 6, 1863, in Chase Papers, Library of

Congress; see also Clay to Brutus J. Clay, October 9, 1863, in Clay Family Papers, Library of Congress.

5 Clay, *Memoirs,* I, 458–59.

6 *Ibid.*, 459.

7 Wade to Clay, February 3, 1870, in Cassius M. Clay Collection, Lincoln Memorial University; Clay to Palmer, April 1, 1870, in Cassius M. Clay Collection, The Filson Club.

8 Clay, *Memoirs,* I, 541, 542, 549; Order Book #58, Fayette County Circuit Court, p. 87, *Cassius M. Clay* vs. *Mary Jane Clay*, Fayette Circuit Court, File 1732, February 7, 1878.

9 Clay, *Memoirs,* I, 501.

10 Clay to Henry Wilson, March 17, 1871, in Henry Wilson Papers; Clay, *Memoirs,* I, 501–2.

11 Clay's speech at Lexington, Kentucky, July 4, 1871, *ibid.*, 502, 503.

12 Clay to Missouri Liberal Republican Convention, January 20, 1872, copy in Cassius M. Clay Collection, The Filson Club.

13 Clay, *Memoirs,* I, 504–7.

14 Speech of Cassius M. Clay in Covington, Kentucky, April 30, 1872, clipping in Cassius M. Clay Collection, The Filson Club; Clay to New York Central Committee, Liberal Republican Party, May 22, 1872, in Cassius M. Clay Collection, The Filson Club; Ethan Allen to Clay, May 18, 1872, Cassius M. Clay Collection, Lincoln Memorial University; Manifesto of Cassius M. Clay, Chairman Provisional Executive Committee of Kentucky, June 18, 1872, in Cassius M. Clay Collection, The Filson Club; Clay, *Memoirs,* I, 508–9.

15 Speech of Cassius M. Clay at Cincinnati, July 7, 1872, from *Cincinnati Commercial*, clipping in the Cassius M. Clay Collection, The Filson Club.

16 Clay, *Memoirs,* I, 507. "To this day many Democrats all over the Union regard the Greeley movement as a mistake; when every man of reflection sees that such was the only road out of the pit of impotency and despair, where the 'Lost Cause' had sunk them."—*ibid.*, 510.

17 Interview with Cassius M. Clay, by a reporter of the *Cincinnati Commercial*, in *Kentucky Register* (Richmond), July 30, 1875.

18 *Kentucky Register*, July 30, 1875; F. W. Johnston to Cassius M. Clay, Jr., October 9, 1875, in Brutus J. Clay Papers.

19 *Kentucky Register*, August 13, 1875, September 24, October 8, 1875. Quoted passage is from speech of August 6, in the issue for August 13, 1875.

20 Clay, *Memoirs,* I, 510–16; *Kentucky Register*, October 8, 1875;

October 29, November 5, 1875; *Memphis Appeal*, October 23, in *Kentucky Register*, October 29; Greenville *Times* (Miss.), in *Kentucky Register*, November 26, 1875; W. A. Percy to Clay, November 17, 1875, from Greenville (Miss.) in Cassius M. Clay Collection, Lincoln Memorial University; Clay to Warmouth, October 10, 1875, in Warmouth Papers, Folder #76, in Southern History Collection, University of North Carolina, Chapel Hill; C. A. Dana to Clay, September 2, 1875; George W. Julian to Clay, September 6, 1875; Wendell Phillips to Clay, October 17, 1875; all in Cassius M. Clay Collection, Lincoln Memorial University.

21 J. P. Knott to Clay, November 17, 1876, in Cassius M. Clay Collection, Lincoln Memorial University. C. Vann Woodward, *Reunion and Reaction: The Compromise of 1877 and the End of Reconstruction* (Boston, 1951), describes the extent of the compromise which settled the electoral crisis, of which Cassius Clay was only dimly aware. See Clay, *Memoirs*, I, 518.

22 J. P. Knott to Clay, November 17, 1876, in Cassius M. Clay Collection, Lincoln Memorial University; Clay, *Memoirs*, I, 507.

CHAPTER XVI

1 Clay, *Memoirs*, I, 518.

2 Interview with Clay by reporter of the *Cincinnati Commercial* in *Kentucky Register* (Richmond), July 30, 1875; Louisville *Courier-Journal*, October 4, 1891; Clay, *Memoirs*, I, 19.

3 Thomas D. Clark, *The Kentucky* (New York and Toronto, 1942), p. 293; Clay, *Memoirs*, I, 550; conversation with Miss Helen Bennett of Richmond, Kentucky, a granddaughter of Clay, summer, 1950.

4 Louisville *Courier-Journal*, October 4, 1891; *Speech of Cassius M. Clay, at Frankfort, Ky., from the Capitol Steps, January 10, 1860* (Cincinnati, 1860), p. 20.

5 Archibald W. Campbell, *Cassius Marcellus Clay: A Visit to his home in Kentucky. His peculiar habits and remarkable career—the Peaceful Ending of a Stormy Life* (New York, 1888).

6 *Ibid.*, p. 5; Clay to his daughter Mary, April 28, 1878, in Cassius M. Clay Collection, The Filson Club; Clay, *Memoirs*, I, 554.

7 *Ibid.*, 554–55.

8 *Ibid.*, 543, 555; *Cincinnati Enquirer*, March 29, 1879, cited in Clark, *The Kentucky*, p. 292.

9 Clay, *Memoirs*, I, 555–70; John R. Johnston to Clay, October 23, 1877, Cassius M. Clay Collection, Lincoln Memorial University. There are additional newspaper articles of the White affair in Clay Scrapbook #2, p. 7, in the private collection of Professor J. T. Dorris, Richmond, Kentucky.

10 Clay Scrapbook #2, p. 213, clipping dated January 20, 1877; Clay to Judge R. H. Stanton, December 28, 1877, in Clay Scrapbook #2, p. 189.

11 Clippings in Clay Scrapbook #2, for 1880, pp. 69, 73, 77, 65, 77, 71; *Kentucky Register*, April 16, 23, May 14, August 13, 1880. Quoted passage is from Clay to editor, Louisville *Courier-Journal*, December 1880, in Clay Scrapbook #2, p. 175.

12 Clay to editor, *Philadelphia Telegraph*, December 14, 1884, in Cassius M. Clay Collection, The Filson Club; speech of Clay at Louisville, August 25, 1884, in Clay Scrapbook #2, p. 217; speech of Clay at Jacksonville, Ill., from Jacksonville *Daily Journal*, October 17, 1884, in Clay Scrapbook #2, p. 229; Clay speech at Lockport, N.Y., from Lockport *Journal*, October 31, 1884, in Clay Scrapbook #2, p. 233. See also Clay speeches in Louisville *Bulletin* (Ky.), February 24, 1884, Rochester *Post-Express*, October 25, 1884, and Plattsburg *Morning Telegraph* (N.Y.), October 29, 1884. Clay denounced the Solid South in post-election comments. "The Solid South holds now sixteen States in armed subjection by suppressing the Republican ballots. In my opinion Mississippi, Louisiana, and South Carolina are as much Republican as Vermont, but they will be probably counted for the Democratic candidate for President. The Republican party in great magnanimity have looked upon these rebel aggressions, by which not only the eleven original seceding States have been united in the Solid South but five Union States also by murder added to the confederates. Now these methods are sought to be unblushingly introduced into the other, the northern half of the Union, to sink our Republic into a confirmed despotism, where the voice of the people shall be heard no more. . . ."—Clay to editor, Albany *Morning Express* (N.Y.), November 8, 1884.

13 Clay to editor, Albany *Sun* (N.Y.), November 16, 1884, in Cassius M. Clay Collection, The Filson Club.

14 Clay, *Memoirs*, I, v, 518–19. All the letters in the Cassius M. Clay Collection at Lincoln Memorial University bear the inscription, in a shaky hand, "Passed. C. 1884." For Clay's difficulty in locating a publisher, see C. A. Dana to Clay, June 9, 1884, in Cassius M. Clay Collection, Lincoln Memorial University.

15 An adulatory review of the *Memoirs*, by A. C. Quisenberry, appeared in the Lexington *Transcript*, August 12, 1886. A clipping of it is in the R. T. Durrett MSS, The Filson Club.

16 Clay's scrapbooks contained numerous articles on agricultural matters; see, for example, the article of April, 1880, on "Seeds," in Clay Scrapbook #2; Allen T. Rice, editor of *North American Review*, February 1, 1886, receipts for article on "Race and the Solid South," in the February issue. There is a MS lecture on "Labor and Capital," 1886, in Cassius M. Clay Collection, The Filson Club. The paper on "Money," read March 4, 1890, to The Filson Club, and an essay for the New York *Independent*, April 25, 1889, are both in Brutus J. Clay Papers. Clay's speech before the Ohio Mexican War Veterans, May 8, 1890, is in Cassius M. Clay Collection, The Filson Club, and Clay's petition of November 13, 1890, to the Kentucky legislature is in the Brutus J. Clay Papers. George W. Curtis to Clay, January 21, 1891, in Cassius M. Clay Collection, Lincoln Memorial University, indicates that Clay advocated the socialization of the railroads and their operation by the civil service, like the post offices.

17 Mrs. Leonora Bergman to author, summer, 1951; Marriage Book Number 24, Madison County, Kentucky, p. 7, Marriage Bond, C. M. Clay and Dora Richardson, November 9, 1894; Clark, *The Kentucky*, p. 293. Launey drifted West and died in a midwestern state in the early 1920's.

18 Clark, *The Kentucky*, pp. 293–94.

19 *Ibid.*, p. 293; Clay to President William G. Frost of Berea College, September 19, 1896, in Berea College Library, Berea, Ky.; *C. M. Clay* vs. *Dora Clay*, Madison County Circuit Court, File 321, Bundle 643, filed August 17, 1898.

20 Clay to the press, March 1898, in New York *Journal*, March 21; printed in Clark, *The Kentucky*, p. 295.

21 Lexington *Herald*, September 11, 1950, tells of the young bodyguard, hired in the spring of 1898. The article also illustrates the extent of the Clay legend. Even the snakes on the White Hall estate were of folklorish monstrosity. One, said the bodyguard, was so big it broke down a fence by slithering over it. Clark, *The Kentucky*, p. 294; conversation with Mr. Charles R. Staples of Lexington, Ky., summer, 1950. *Richmond Climax* (Ky.), July 8, 1903. Dora died in 1914 at the age of 35. She had been married five times. Lexington *Herald*, February 14, 1914.

22 Conversation with Mr. Charles T. Dudley of Richmond, Ky., July 18, 1950; with Mrs. Jane Clay of Richmond, summer, 1950;

and with Warfield C. Bennett, a grandson of Cassius Clay, also of Richmond.

23 H. C. Howard to Cassius M. Clay, Jr., a nephew of the old general, April 14, 1899; Brutus J. Clay, a son of General Clay, to Cassius M. Clay, Jr., August 21, 1899; and Clay to Cassius M. Clay, Jr., October 1, 1899, all in Brutus J. Clay Papers; Lexington *Morning Herald*, April 6, 1901.

24 Mr. Charles T. Dudley to author, summer, 1950; Lexington *Morning Herald*, April 6, 1901; *Mary B. Clay* vs. *C. M. Clay*, Madison County Circuit Court, Box 352, Bundle 736, filed April 11, 1901, has the notation, "Executed upon Cassius M. Clay by leaving a true copy of subpoena in residence of Cassius M. Clay—he being barred up in his residence and refusing admittance or to come out and be notified of my business with him." *Madison National Bank* vs. *C. M. Clay*, Madison County Circuit Court, Box 353, Bundle 707, filed November 22, 1901, has a similar note: "Executed November 22, 1901, by having a true copy hereof delivered at the residence and at the feet of Cassius M. Clay, he refusing to permit officers to his house by threatening to kill any person entering his house without his permission." *H. H. Collyer, Sheriff*, vs. *C. M. Clay*, Madison County Circuit Court, File 256, Bundle 713, filed February 26, 1902. *John F. Wagers, Sheriff*, vs. *C. M. Clay, James Bennett, Mrs. James Bennett, Mrs. Annie Crenshaw* [Clay's daughters and son-in-law], Madison County Circuit Court, Box 362, Bundle 724, filed January 13, 1903. Clay received an annuity of $360 from each of his four daughters. On one occasion his grandson Warfield Bennett went out to White Hall to deliver a check for his mother's payment. Clay refused it because he would have to endorse it. "Long ago I decided I would no longer sign my name to anything," he told the boy.

High Sheriff Josiah Simmons' letter is quoted by William H. Townsend to the Lincoln Group in Washington, D.C., in Washington *Evening Star*, November 17, 1954.

25 Deposition of W. O. Bullock, M.D., October 10, 1951, in the private collection of J. Winston Coleman, Jr., Lexington, Ky.

26 *Richmond Climax*, July 22, 1903; Clark, *The Kentucky*, pp. 272–73.

27 *Richmond Climax*, July 29, 1903; Louisville *Times*, July 25, 1903; Clark, *The Kentucky*, p. 273; J. Winston Coleman, Jr., *Last Days, Death, and Funeral of Henry Clay*, p. 30, n. 56.

A NOTE ON THE SOURCES

Materials for a study of the career of Cassius M. Clay appear throughout the papers and memoirs of nearly all the leading Republicans and antislavery politicians of his time. He readily collected friends and enemies, he was involved in many conflicts of both local and national character, and he traveled and spoke widely over the country. All this contributed to the record of his life. Clay saved his correspondence and maintained scrapbooks, but much of this valuable source material was scattered or destroyed. Important collections of his papers, however, may be found in the Brutus J. Clay Papers, in the private collection of Cassius M. Clay of Paris, Kentucky; in the Cassius M. Clay Collection, The Filson Club, Louisville, Kentucky; and in the Cassius M. Clay Collection, Lincoln Memorial University, Harrogate, Tennessee.

Other Clay letters are in the Breckinridge Family Papers, the Salmon P. Chase Papers, the Hamilton Fish Papers, the Giddings-Julian Papers, the Robert Todd Lincoln Collection, the Edwin M. Stanton Papers, the Elihu B. Washburne Papers, and the Henry Wilson Papers, all in the Manuscripts Division of the Library of Congress. In addition, there are Clay manuscripts in the John A. Andrew Papers in the Massachusetts Historical Society, the Salmon P. Chase Papers in the Historical Society of Pennsylvania, and in the William A. Seward Papers and the Thurlow Weed Collection in the University of Rochester Library. All of these collections contain personal letters which provide insight into the private purposes and ambitions of the man Clay.

As a newspaper editor, orator, and pamphleteer, Clay frequently expressed the political and economic program he advocated for his state, the South, and the nation. Among his printed writings, the most valuable are *The Life of Cassius Marcellus Clay: Memoirs, Writ-*

ings, and Speeches (Cincinnati, 1886), *The Writings of Cassius Marcellus Clay*, edited by Horace Greeley (New York, 1848), and the pamphlets containing his important public addresses. Much of his antislavery opinion appeared in the columns of *The True American*, a newspaper he edited in Lexington, Kentucky, in 1845 and 1846. Other newspapers in Lexington, Frankfort, and Cincinnati, as well as the national press, carried articles and speeches by and about Clay.

Memoirs, diaries, and printed correspondence of many of his contemporaries contain revealing views of Cassius Clay. Valuable among these are the works of John G. Fee, Thomas F. Marshall, John Bigelow, Salmon P. Chase, John J. Crittenden, James G. Birney, George W. Julian, Jane Grey Swisshelm, E. D. Mansfield, Abraham Lincoln, Orville H. Browning, Jeremiah Curtin, William H. Seward, John Hay, Benjamin Moran, Thurlow Weed, and Gideon Welles.

Beyond these primary sources, many monographs and special studies offer important information on Clay's career. William E. Baringer, *A House Dividing: Lincoln as President-Elect* (Springfield, 1945), records Clay's frantic efforts to get into Lincoln's cabinet. Ruhl J. Bartlett, *John C. Fremont and the Republican Party* (The Ohio State University Studies . . . No. 13, Columbus, 1930), mentions Clay's services in the campaign of 1856. Mary Rogers Clay, in *The Clay Family* (Louisville, 1899), gives a genealogy of Cassius Clay's family. J. Winston Coleman, Jr., in *Slavery Times in Kentucky* (Chapel Hill, 1940), and in a pamphlet, *Last Days, Death, and Funeral of Henry Clay* (Lexington, 1951), tells of Clay's antislavery activities. In the latter work, Mr. Coleman records the electrical storm over Lexington at the time of Clay's death. A fair appraisal of Clay's influence upon the embryonic Republican party is in Andrew W. Crandall, *The Early History of the Republican Party* (Boston, 1930). Jean M. Howard's unpublished M.A. thesis at the University of Kentucky, "The Ante-Bellum Career of Cassius Marcellus Clay" (1947), is a partial study of the origins of Clay's career, based largely upon his *Memoirs*. Asa Earl Martin, in *The Anti-Slavery Movement in Kentucky Prior to 1850* (Louisville, 1918), provides a useful survey of the work in which Clay was influential. James Rood Robertson's *A Kentuckian at the Court of the Tsars: The Ministry of Cassius M. Clay to Russia* (Berea College, Ky., 1935) is an interesting account of Clay's diplomatic efforts, based upon State Department archives. Albert A. Woldman gives a more general account of American-Russian relations during the Civil War in *Lincoln and the Russians* (Cleveland and New York, 1952).

INDEX